NINPO

LIVING AND THINKING AS A WARRIOR

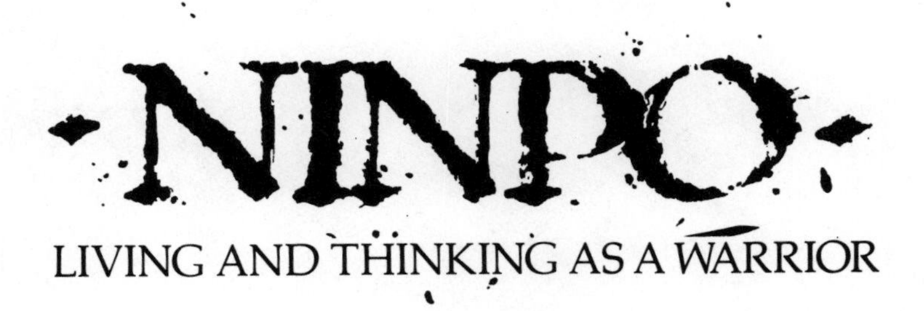

NINPO

LIVING AND THINKING AS A WARRIOR

JACK HOBAN

CB
CONTEMPORARY BOOKS
CHICAGO

Library of Congress Cataloging-in-Publication Data

Hoban, Jack.
Ninpo : living and thinking as a warrior / Jack Hoban.
p. cm.
ISBN 0-8092-4725-9 (pbk.)
1. Ninjutsu—Philosophy. I. Title.
U167.5.H3H627 1988
355.5′48′01—dc19 88-20357
CIP

Published by Contemporary Books, Inc.
180 North Michigan Avenue, Chicago, Illinois 60601
Manufactured in the United States of America
Library of Congress Catalog Card Number: 88-20357
International Standard Book Number: 0-8092-4725-9

CONTENTS

ACKNOWLEDGMENTS

The following persons were instrumental, either directly or indirectly, in the writing of this book: Dr. Masaaki Hatsumi, Dr. Robert L. Humphrey, Stephen K. Hayes, and my editor, Donavan Vicha. I'm sure, however, that they all don't agree with everything I will say herein.

Special thanks to Gregory Manchess, Tom Young, Cindy Lewis, and Ken Lux, for sharing their arts with me.

This book is dedicated to Dennis W. Moran.

INTRODUCTION

I believe that a whole philosophy, like a whole person, should have a mind, a heart, and a body. The word for "whole person" in Japanese is *tatsujin*. Although I am not Japanese, I enjoy thinking about the concept that that word embodies: Completeness. This is what I have come to understand as the goal of my journey down the warrior path. Of course I am not there. Yet. In fact, I have times when I doubt whether I am even far enough along to write a book such as this. But then, I think: If I waited for enlightenment before I began to write, this book would probably never be written. And who would understand it if I did write it in my enlightened state?

Therefore, I offer this volume as a glimpse at a process, albeit an incomplete one. I invite you to grow with me.

There is no ninja philosophy that I have ever been taught, word for word, that approaches what I have discovered for myself through immersion in this growth process.

The discussions in this book are based on a rational epistemology. In other words, I proceed from the premise that there is such a thing as reality, and that man is capable of acquiring knowledge of that reality by a process of reason. Therefore, I believe that my philosophy is rational and usable in the real world. However, I must outline the process that has lead me to adopt my view of man's metaphysical nature if I am to justify morally my involvement in the warrior arts. Although I no longer need to justify this involvement to myself, the reader is strongly urged to examine critically every statement that I make. One of the eventualities of living as a warrior may include the taking of a human life. Therefore, these things *must* be thought about very, very deeply. Without a true moral sanction to act, one can never be a warrior—only a killer.

SOME PREMISES AND DEFINITIONS

Observe this persistent trend: The thinking man is unable to act, and the man of action does so without thinking. As a result, the ability to understand and express man's most important value, love, is severely impaired for both of them. One "feels" love, but cannot make a connection. The other consummates his "connections," but they are connections without love. By love I mean the full appreciation of the value of another's life to your own.

Values imply choices. In order to choose, you must be able to think and commit. Love implies a rational act of commitment to a value system. If you cannot act, you cannot participate in the value-for-value relationship that is true love. Though love is usually termed a "feeling," it is also a thinking and doing affair.

As this is a book on warriorship, it is important to define the term as it will be used herein. *A warrior is a man of action, guided by reason and motivated by love.* Obviously a discussion of warriorship must contain the usual aspects of warfare, strategy, and fighting. Yet, what does a warrior do when he is not in a war? He continues to live, of course. How? He lives according to the same set of moral precepts that he uses when engaged in warfare. In this book I will posit that the only moral application of warrior skills is self-protection. Therefore, a warrior is a person who can and will act in defense of himself or others of his choice, regardless of the scale or scope of the situation.

To understand that which is a truly self-defensive action requires two things: a value system, and the ability to reason. In other words, a warrior must know why and when it is moral to act, and why and when it would be immoral to use his skills.

To choose to defend oneself, one must understand the value of oneself. To choose to defend another implies that there is something of value to defend. To act effectively, certain skills are required. This book is about these three things: reason, values, and action.

The philosophy in this book is mine, and mine alone. Of course I have been profoundly affected by many persons and ideas. As a tribute, therefore, I would say that the mind of my philosophy is essentially Aristotelian with help from Thomas Aquinas and Ayn Rand, although I believe their philosophies to be dangerously incomplete without certain balancing disciplines. The heart of my philosophy is basically the dual-life theory, as described by Dr. Robert L. Humphrey, although we have several important differences in emphases as well. The physical aspects of my philosophy are gleaned from the ninpo taijutsu training method of Dr. Masaaki Hatsumi and his students, too numerous to mention, who are helping me understand it.

I know that my evolving philosophy would be more incomplete if I did not have the opportunity to read and, in some cases, speak with these other philosophers. But, make no mistake about it, this is a book on warrior philosophy. The reader should keep that in mind. I am not an intellectual; the first section of this book is no more than a short explanation of my personal political views. Nor do I possess the insight and love of my fellow man that would qualify me as a humanitarian; the second section of the book is a rather simplified moral philosophy. I have tried to be more complete in the third section, however, because what I am truly inclined to is warriorship. I have always striven to attain the tools to do the moral thing, regardless of the consequences. In this world, those tools must be formidable. From my study of the lives, philosophies, and methods of the Japanese sect known as *ninja,* I have come to believe that the study of *ninpo* provides the tools to live a moral life on this earth. That is why I have adopted their training methods into my life ways.

NINJA WARRIORSHIP

Obviously I am not a Japanese ninja. Western critics might well wonder why I am so influenced by this particular group of people. In the last thousand years, the ninja have been far from the dominant culture in Japan, much less anywhere else. If anything, they were a counterculture. The history books are written by the dominant societies—in this case the samurai, so the story of the ninja is not readily available. We may never know the full story. For most intents and purposes, the ninja were a vanquished people. But their legacy remains. Their legacy is what is important.

It has been barely twenty years since the first Westerner glimpsed the hidden world of the ninja that had been a secret for more than eight centuries. This world had shrunk to the size of a small room in the back of a bone doctor's clinic in Noda City, Japan. Since then, this small-town doctor and his art have become the most talked about, written about, and controversial of any single martial tradition in memory. The doctor's name is Masaaki Hatsumi. He is a ninja.

The reason for all of the excitement, perhaps, is the recent exposure of Westerners, in sensationalized movies and novels, to certain legends that have been passed down through the centuries in Japan. These folktales describe the ninja as sorcerers and demons, the ultimate warriors who could win a battle by the power of intention alone. A ninja was reported to be able to cloud the mind of an enemy, walk on water, and even fly.

I believe that more responsible and historical accounts of the ninja have cleared up the popular exaggerations, but there still remains a fascination for this "shadow warrior" and the art of ninjutsu.

Ninjutsu is usually translated as the "art of stealth." The Japanese character, "nin" or "shinobi," that stands for ninjutsu has many nuances of meaning, including perseverance, endurance, and sufferance. It is most commonly used to refer to the specific methods used by the ninja. My favorite translation of ninjutsu is "the art of subtlety."

The art was born in central Honshu (the largest of the Japanese islands) about eleven hundred years ago. It was practiced by mountain families in an area not unlike the American Appalachians. The terrain was rugged and remote. The "art" was never really established as such, but developed over the course of many years. The crystallization of various ideas into a definitive lore was a process that took several generations before the specific name ninjutsu came into use.

The families that adhered to this lifestyle were great observers of nature. They also were deeply inspired by several spiritual disciplines that became interwoven into their philosophical method.

One of them was *Shinto,* or the way of the kami. Kami is the general Japanese word for god or deity. It implies, however, a feeling for a sacred or charismatic force. The early Japanese regarded their whole world: the rivers, mountains, lakes, and trees, to be imbued with spirits. Although ancestor worship and a filial love of everything Japanese were later tenets of Shinto as an organized native religion, Shintoism was originally a form of nature worship.

The second influence on the ninja was a unique spiritual discipline called *Mikkyo.* These original Mikkyo practices are far removed from the *Tendai* or *Shingon* mikkyo religions that are practiced today in Japan and elsewhere. However, they did spring from the same sources in Tibet and China. Mikkyo, for the ninja, was less a religion than a method for enhancing personal power. The methods included the use of charged words and symbols to channel one's energies and intentions toward specific goals.

It is generally accepted that these methods originated outside of Japan. After the fall of the T'ang dynasty in China, many outcast warriors, philosophers, and strategists escaped to Japan to avoid punishment at the hands of the new rulers. It is believed that ninja families were exposed to many of these exiles' sophisticated warrior strategies and esoteric philosophies over the centuries. Mikkyo must certainly be included as one of these imported influences.

The ninja are acknowledged to have been very much influenced by a cult of mountain ascetics called *Shugenja,* who roamed the same mountainous sections. The *Shugendo* method of spiritual self-discovery consisted of subjecting oneself to the harsh weather and terrain of the area in order to draw strength from the earth itself. They would walk through fire, stand beneath freezing waterfalls, and hang over the edges of cliffs in an effort to overcome fear and assume the powers of nature.

It would not be correct to say that these three spiritual methods were the actual roots of ninpo, but there is little doubt that they were concurrent influences for the ninja families. Ninpo was and is a separate philosophy.

The ninja were not particularly warlike, yet they found themselves in the position of being harassed oftimes by the ruling society of Japan. They were routinely subjected to unfair taxation and religious persecution. The ninja eventually learned to act more and more efficiently in their own self-defense. They used their superior knowledge of the workings of nature, as well as specific military techniques passed down through the years, as weapons against the numerically superior government armies. They used any ruse, harbored any superstition, and employed any strategy to protect themselves. If necessary, they would use devious political manipulations to ensure peace.

There were as many as seventy or eighty ninja clans

operating in the Koga and Iga regions of Japan during the height of ninja activity. During certain historical periods they were minor powers in and of themselves. But most of these ninja were descendants of, or were themselves, displaced samurai. Therefore, they operated at the periphery of the political machinations of the government. Sometimes a family would use its military or information-gathering resources to protect its members from becoming pawns in a power play between competing samurai factions. Occasionally, a ninja family would support one faction over another, if they felt it to be to their advantage.

As with any society, there were renegades who misused the training they received. Occasionally, "ninja" would rent themselves out for espionage or assassination work. These outcasts have become the stereotype of the skulking ninja that we see today in the media. They were an aberrant minority. The average ninja worked very much in confluence with his family and community goals.

Ninja were not always primarily soldiers. Of course, certain ninja operatives, or *genin*, were trained from childhood as warriors. But this training was usually precautionary. Genin ninja knew that they might be called to help protect the community at some future time, but, they often spent most of their lives as farmers or tradespeople. Ninja intelligence gatherers sent to live in the strongholds of potential enemies were rarely required to act overtly.

If an operative was called to action it was as a result of a carefully plotted, and usually desperate, plan. The genin would be contacted and assigned a mission by his *chunin* superior. The chunin, or middle man, was an intermediary between the *jonin* family leader and the operative. Jonin made all philosophical and long-range strategic decisions for the clan. Often, the identity of the jonin was kept secret from chunin and genin, alike. Of course certain historical periods required more secret activity than others.

Eventually this activity virtually died out altogether but the legacies, in some cases, remained.

HISTORY OF THE TOGAKURE RYU

The *Togakure Ryu* (tradition) of ninjutsu is the senior of nine martial ryu for which Masaaki Hatsumi holds the *maki-mono* (secret scroll legacies). The keeper of the maki-mono is recognized as *soke* or spiritual leader of that particular tradition.

Although historical records are incomplete and imprecise, it is believed that Togakure ryu ninjutsu was founded by Daisuke Nishina in the 1100s. Daisuke, himself, was exposed to Shugendo training as a young man in the Togakure (now Togakushi) region close to present day Nagano. Daisuke's proficiency in the military arts received the ultimate test in 1181, when he joined with a local leader named Kiso Yoshinaka to battle Heike troops sent to the area to subdue the populace. After three years of fighting, the local resistance movement was crushed and Daisuke was forced to flee for his life.

He wandered south and took refuge in the Iga province outside Kyoto. There he met Kain Doshi, a *yamabushi* warrior monk, who indoctrinated him into the spiritual perspective that rounded out his warrior education. Daisuke, in celebration of the birth of his ultimate power as a warrior, assumed the name of his homeland. He was thence known as Daisuke Togakure; and his descendants are known as Togakure ryu ninja. The ryu flourished for several centuries and experienced its greatest time of power in the 1500s before the rise of Iyeyasu Tokugawa.

The Tokugawa Shogunate brought to Japan a time of unparalleled peace and tranquility. The life-or-death environment under which the ninja families had operated for many years was eased. Most warrior training was abolished or ritualized; and many warrior traditions simply died out for reasons of disuse or lack of interest. Still, a number of these ninja systems remained alive, passed on through the centuries by a few special persons who knew that these methods should never be totally forgotten.

There are ninja still today, therefore, though in an

evolved sense. Surprisingly, there are more active practitioners, worldwide, than there have ever been. Dr. Hatsumi feels a responsibility to make the methods of ninpo available to others, all over the world. He has spoken of a feeling that his teacher, Toshitsugu Takamatsu, still speaks to him from the grave, instructing him to carry on the ninja traditions.

The reader may wonder why this man feels so strongly about going public with this knowledge after so many years of obscurity. Does the current world situation point to a need for these methods? I cannot say for sure.

Those persons who have been exposed to Dr. Hatsumi and his students who are now teachers, are less preoccupied with the tales of ninja's shrouded past than with the application of ninjutsu principles here and now. In order to understand how this might be, it is important to take a short look at how the philosophy of ninpo was formed.

THE ROOTS OF NINPO

Although there has been an evolution of ninpo as a life philosophy over the centuries, the fundamental principles have remained virtually unchanged. Togakure ryu ninjutsu is more than 800 years old. Except for a relatively short period of notoriety prior to the reign of the Tokugawas, the art lived quietly in the hearts of just a few overlooked souls. The ninja were a separate society from the urban centered ruling class and the nonprivileged classes which served them. Consider the gulf that must have existed between the new American government and the American Indians during the first 125 years following the signing of the Declaration of Independence. Although this is an incomplete and potentially misleading analogy, it gives the reader a better perspective on how ninpo may have developed as a counterculture to the samurai-dominated Japanese society.

For hundreds of years those ninja families lived in the

mountains, practicing their esoteric methods of approaching enlightenment through gaining an understanding of the fundamental laws of nature. History had taught them that they must be prepared to protect the sanctity of their endeavors. Toward that end, they perfected a system of combat arts that has earned them the reputation for being the most amazing warriors the world has ever known. It is this reputation that initially attracts most enthusiasts.

The ninja's reputation is put into a better perspective when some facts are brought to light. First, ninja were not wizards or witches, of course, but ordinary men of a unique and misunderstood philosophical viewpoint. This philosophy became manifest in their combat method. Hence, we refer to our art as ninpo, the "*po*" suggesting "a higher order," or "encompassing philosophy." The samurai approach to combat was called bushido; it evolved from a general set of guidelines for the gentleman warrior into a stylized and formal discipline. The ninja, though ostensibly holding many of the same values as the original samurai, never developed such dogma.

Their sometimes devious tactics were seen as cowardly and disgusting.* From the ninja point of view, however, guerrilla warfare versus a numerically superior force was plain good sense. The ninja were outnumbered, as a rule, so they had to use unusual methods if they ever hoped to survive. Nevertheless, victory was not always ensured. Japanese history books are dotted with instances of entire ninja clans being destroyed.

Many times, however, the unusual methods did succeed. Without a working knowledge of the ninja philosophy, their contenders were unable to unravel the ninjutsu strategies. The ninja only *seemed* like wizards.

---

*During the American Revolution, British Red Coats, accustomed to marching to battle in orderly phalanxes, were decimated by camouflaged Green Mountain Boy guerrilla forces shooting from behind trees. The Red Coats must have felt the same way. America won the war, however, so our history books do not stress the British side of the story.

Second, the chronicles of certain notorious exploits, which have created the prevailing image of the ninja as conscienceless criminals, were written after the fact by historians who were sympathetic to the samurai point of view. Since ninja were not bushi (adherents to the samurais' strict code of martial ethics) they were looked down upon as being rather uncivilized to begin with. Historical data are always tainted with the prejudices of the predominant society of the particular era.*

Third, the exaggerations of ninja abilities were fostered by the ninja themselves as a deterrent to outside interference. The demonstrated prowess of the ninja as warriors, as well as the fact that they were such a closed and uncommunicative society, combined to create an opportunity for them to exaggerate their own skills and surround themselves with an eerie cloak of intrigue. It is more reasonable to be frightened of something that is not understood than by something that has an easily understood potential. Thus, this frightening and supernatural mystique was born.**

If, however, it was merely the guerrilla tactics of the ninja that were useful, I doubt that the lore of the ninja would be of interest to the wide range of people who enjoy

*For example, it was easy for the early white historians to depict the American Indians as scalping savages. Later, more reliable research uncovered evidence that it may have been the French who instituted the practice of scalping so that bounty hunters could not exaggerate how many Indians that they had killed, and so be overpaid. Even though accounts more sympathetic to the plight of the Indian tribes at the hands of the whites have been written, the prejudices are deeply ingrained and may never be overcome. The same may happen to the ninja.

**In Viet Nam, many American soldiers were "spooked" by the thick, black jungles of Southeast Asia, and an enemy that was everywhere, yet never there. Thus the Viet Cong were able to use guerrilla tactics with great success. The ninja, over the course of many centuries, made an art out of preying on the irrational fears and superstitions of their enemies.

practicing ninjutsu today. Far more than an arcane set of stealth or assassination techniques, ninpo, or the essence of the ninja's outlook, is a physical, emotional, and spiritual method of self-protection from the dangers that confront those on the warrior path to enlightenment.

Ninpo taijutsu is the physical training method for opening up oneself for this kind of growth. By understanding the natural body, one gains insight into the natural universe. But this process is, of necessity, a very personal affair. The training method merely illuminates certain presituational fundamentals. Each person on the warrior path must fill in their own blanks. That is why my philosophy, as each warrior's, is unique.

The training method of ninpo taijutsu allows one to become reacquainted with the natural laws of the universe. This insight can then be used as a foundation for the growth of the mind and heart. Without this presituational "starting point," it is easy to violate the laws of nature through pragmatism or uncontrolled passion.

The most visible aspects of ninpo, of course, are the physical skills. The word taijutsu is almost a generic concept. Literally it means "body techniques," or, how to move your body. Obviously each person has his or her own way of moving. How well or not this movement conforms to natural laws, as well as the spirit of shinobi, is the issue addressed by ninpo. In so far as the movement is natural, the action is good. In so far as the movement is unnatural, the action is bad.

If we expand this concept we might see that natural movement can work to expand our lives, unnatural movement to restrict us. Ninpo taijutsu denotes freedom of movement, in the sense that it is natural and liberating, while remaining confluent with the philosophy of ninpo. If I am forgiven at the start for using the Japanese word taijutsu as a metaphor for freedom, some of the nuances of my philosophy will be easier to comprehend later on. That we should be free is a subject that permeates every line of this book.

Ninpo is an admonition to move subtlely, as nature does, to ensure that your actions do indeed blend with the universal scheme. This subtlety will also keep the practitioner safe from those persons who live by restricting the freedom of others.

WARRIORSHIP

I cringe when I hear the words "warrior" and "enlightenment" bandied around as much as they are these days. Most people, I am sure, yawn at the mere mention of them. Discussions of these subjects have deteriorated into intellectual and semantic fencing; and, as with any sport, rules have been invented to make this kind of philosophizing more fun. Unfortunately, these rules allow the suspension of reason and ignore the laws of nature. Sincere seekers of knowledge may be tired of the game. They begin to ask themselves whether there is such a thing as a foolproof method to ensure enlightenment.

The more naive (or irrational) continue to search in vain for someone to teach them the secrets of the universe. I cannot help them. But, as mentioned before, there is always room for growth, and there are methods that contribute to that growth. One can become a part of that process through training. My unique view of the warrior has developed over years of military and martial arts training, and later, through academic and philosophical studies. I see the personage of the warrior as one who is self-reliant and responsible for his own actions. Warriorship is a lifestyle of seeking answers to many questions, perhaps the same questions that were asked by the original ninja as well.

Ultimately, to be a warrior is to be a person of action. That is fundamentally what this book is about: Becoming a person who can act. What do you do with this ability? Ninpo admonishes us to act to defend your own or someone else's freedom. To be free and to assume the right to act, however, one must assume the corresponding responsibilities.

One must know what to do.

One must know when to do it.

One must know why.

I am aware of the fact that I may be criticized for drawing so heavily from a small culture to expound what I feel to be a philosophy that may have universal application. I cannot apologize for that. It merely proves how a valuable idea, even if it is used by a limited number of people, can result in great good.

The reader should acknowledge, also, that the ninja's action-oriented philosophy has survived for close to nine centuries. It has truly withstood the test of time.

I believe that ninjutsu has found popularity in the western world because it offers more than a new look at basic martial arts techniques, or added a new twist to sport-oriented martial arts competition. Rather, ninpo provides the serious practitioner with a total physical, mental, and spiritual life philosophy. What I offer in this book is my interpretation of the ninpo code of behavior—a manifesto, if you will—for the warrior.

I present this code in three parts. In the first part I deal with the intellectual issues that form a basis for this philosophy, and through economic and political examples, explain the ethical foundation of the practice of ninpo. The mind of the warrior requires intellectual training just as much as the body requires physical training.

The second part of this book addresses the spirit of the warrior, and in so doing, explores the phenomenon of love. Love is the source of power that clarifies the mind and sanctions the actions of the body. Again, brief examples from our society's economic and political framework will be used in interpreting how love becomes manifested in the life of the warrior.

The physical training of the body will be introduced in the third part of this book. While it will, by necessity, concern itself with practical techniques and body movements, such training is meaningless, if the body's actions are not born from an ethical mind and a loving spirit.

I was inspired to adopt these ideas as my own and carry on with them in my own way. I have no way of knowing whether those other philosophers who have inspired the ideas in this book would approve or disapprove of what I have done. Therefore, you as the reader, can approach this book in any way that you like. If you need to understand the reasons behind why I believe that this art is so important to today's world, start at part one. If you are curious about the motives and motivations behind this philosophy, start at part two. But, if you want to learn the power of this art, start at part three.

It is truly up to you.

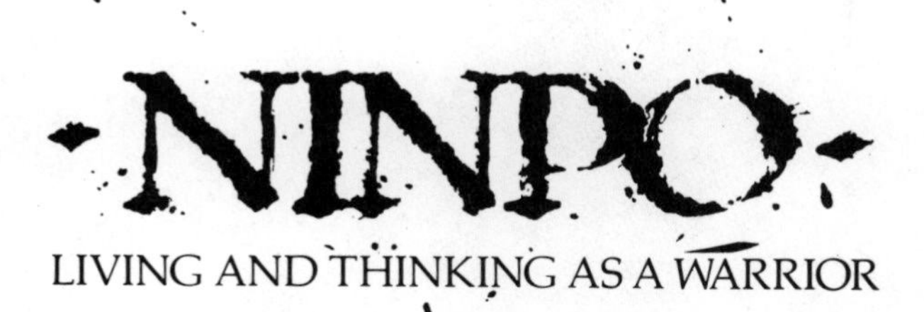
NINPO
LIVING AND THINKING AS A WARRIOR

PART ONE
THE MIND OF THE WARRIOR

A human being must use his brain to live. If he doesn't use his own, and he is alive, he is using, or is dependent on, someone else's.

Human beings are very different from other animals in this respect: We have a free will. Our animal instincts are evolved more into the form of tendencies. Our free will can override these tendencies, for good or bad.

In many martial traditions, animals are used as metaphors for certain strategies and forms. In our own Ninpo tradition, certain techniques are named after animals and their characteristics. Also, martial art names, such as moko no tora (Mongolian tiger, Toshitsugu Takamatsu's warrior name), were given to senior figures. The tradition continues to this day. But these names are inspirational only. It is far better to learn to fight with the full power of the human being than any animal. This will remain true until the unlikely time when Mongolian tigers are looking at us in zoos instead of the other way around. How is it that we can neutralize a beast as powerful as a tiger when we are so much physically weaker? The answer involves the use of our rational brain.

Aristotle uttered one sentence that has changed man's metaphysical view of himself forever. He said: A is A, a thing is itself.

Regardless of whether we like it or not, certain things exist, and other things do not exist. Hence we have reality. Life is not an illusion, it is real. That is why our rational brain is so powerful, it is perfectly designed to function in such an environment: The real world.

There are those who would have you believe that life is an illusion, that no one can be sure of anything, that things may make sense in theory but don't work in practice. Whether they know it or not, these people have as their objective the negation of man's mind. They would never admit it, but they are out to rob you of your most powerful tool to live on this earth. They may never choose to acknowledge it, but without your rational brain, you would die.

No matter what semantics we may be led to play by those who deny the necessity of using one's brain, we use our brains quite rationally much of the time. We *have* to. Food is not an illusion, shelter is not an illusion, clothing is not an illusion. Without them we would die. Many people have illusions about death, but no one can prove them with the organ that keeps us alive: The thinking, rational brain.

We all try to turn our brain off, at least every once in a while. It is amusing to attempt a suspension of our indissoluble marriage with reality. This is not inherently harmful. We are allowed to have fun. Since we have a free will, however, we can use that creative, free-thinking ability of the brain to live in the real world, or make believe that there is no such thing.

We can forge new realities or conjure up empty illusions.

We can create a skyscraper or a pink elephant.

We can plant a garden or live in candyland.

We can build a fire or dream about mink coats as we freeze to death. It is our choice.

But we seldom allow ourselves to go too far from reality. Why? Because our lives are too real and precious to give up. We don't want to die. And that is what would happen.

Human life is real and has value. A is A. Things that do not promote life are irrational. Things that are life giving are rational. We are very familiar with reality. We live it every day. We must acknowledge it or die. When we live with it, we are the masters of our own destiny. When we don't, we forfeit our lives to anyone who chooses to take it. This is very important to the warrior. He owns his own life.

To live, one must obtain the means to live. One must have those things that sustain life. To live above the base level of existence, one must have a wealth of those things that sustain life. When wealth is obtained and used rationally, it leads to an improvement in the quality of life. Rather than living in a cave, eating only what we can forage or hunt, and remaining subject to the forces of nature, we can live in a warm home, shop at a local store, and use our free time for intellectual and spiritual growth. We have the opportunity to interact with our world as a physical, intellectual, and spiritual being. Of course, we may become alienated from nature and from reason. But the point is, that with wealth, we have the choice. It is important to understand wealth as an abundance of those good things that have, as a rational purpose, the sustenance of life. It is also important to understand how one obtains wealth, and differentiate between moral and immoral methods of acquisition.

There are three ways to obtain wealth. The first is to produce it. The second is to steal it. The third is to appeal or beg for it. To steal, one must steal from someone else who has wealth. To appeal, one must appeal to someone who has wealth. Eventually, someone must produce or there would be nothing to steal or beg. That is reality. There is no such thing as manna (or pennies) from heaven in any but a metaphorical sense.

Since the means to life are life-giving, production is

moral. Since stealing or appealing may result in the wresting of that which is life-giving from another in violation of an inalienable right, to steal or to beg is not life-giving. Thievery and begging are morally wrong. They are acceptable only when they prevent death. The need to steal or beg to prevent death cannot occur in a moral society, where people are free to produce and deterred from stealing and/or begging.

ETHICS

It is not enough to feel that this is true. One must understand the reality of these things, one must know why.

I feel that living in a free society requires a warrior to be a philosopher, as well. In this time of relative peace, it is unnecessary for the warrior to merely be a tool toward some institutional purpose. He must make the everyday decisions that dictate how and when to use his skills. This, I believe, is the form of the current stage of the evolution of ninpo. It is the manner in which the ninja lives in a semi-free society. In order to function he must know why things are as they are. He must know why an act is moral, and why another is immoral.

It may seem that I am equating the word *moral* with the word *productive*. Productivity is an attribute of the moral. A moral person does that which is life-giving. That, of course, includes the production of those goods and services which sustain himself or others, and increase the quality of our lives. We all require sustenance to live. Those who provide that sustenance are moral, as long as they do not exploit some people in order to create sustenance for themselves and others.

Ethics, therefore, or the study of morality, is one of the branches of philosophy with which the warrior must be concerned. The other two branches of philosophy: metaphysics, or the study of existence, and epistemology, or the study of how man knows things, are a foundation to the basic calling of the warrior: We are, we can reason, so, what do we do? As rational human beings we have to know *what* to do and *why*. As warriors, we are willing to do it. These three pieces are part of a whole that equals correctness, or, for me, "how to live." Offering the adage: "when in doubt, do something, anything" is a cruel trick to play on the person desperately seeking reason in his life. The adage should be: "when in doubt, think." It is important to have a rational, presituational value system to use as guiding fundamentals to understand situational issues.

As a warrior one must live to uphold moral, life-giving values, and to stand beside man's right to life, liberty, and the pursuit of happiness.

THE MORAL ROLE OF GOVERNMENT

Putting the responsibility for protecting man's life, liberty, and his right to pursue his own happiness on the shoulders of the warrior may seem a bit much. Aren't these functions the role of government?

Governments, in general, are particularly uninclined to teach, much less, enforce, individual rights. That is why the concept of government envisioned by our founding fathers is unique. For the first time, the combined might of the majority was to be used to protect the rights of the individual. That is the proper role of government.

Government should protect the individual from the immoral, those who would deprive a moral, productive individual from what he needs to live as a total human being on this earth.

Can you see how this differs from conferment or bestowal, as if from a position of authority? We assume that man has inalienable rights as an individual. It is not the role of government to give anything to an individual, including his rights (or forcibly redistribute one individual's wealth to another). Governments can only hold his intrinsic rights fast. Ultimately, only an individual can be moral.

The armies of a nation hold its citizens safe from conquest from outside forces. Police act, internally, as a deterrent and as protection against the criminally immoral. A legal system mediates disputes, and proscribes punishment for the criminally or civilly immoral. All other requirements are the responsibility of the individual, unless he and others choose to voluntarily cooperate with each other to accomplish or produce that which no one man can accomplish or produce alone.

The most sensible reason for men to cooperate is to create life-sustaining wealth. The only moral way for a man to become involved in the creation of wealth is through his own volition and according to his own effort and ability. There is a name for this kind of arrangement, although there has never been a pure example of it in the history of this planet. It is called the free market. There is also a name for the lubricant of voluntary cooperation, although it does not exist as a functioning medium in most places on this planet. It is called money.

In a free society the individual should be allowed to produce at the level commensurate with his skill, energy, and ambition. Governments need not and should not interfere with business. The theory of the free market allows for the more-productive person to have a better life, for the less-productive person to have a life that is not as satisfying. Since man is to be a free agent, he can choose what to do, how to do it, and for whom to produce. First he would produce for himself and his loved ones, and then, if he wished, he could make voluntary contributions to those whom he felt deserved his help, such as those who were physically or mentally unable to produce enough to live, such as the handicapped, sick, elderly, mentally impaired, or orphaned.

Since the free-market system is based on the production of life-sustaining wealth, it is a moral system. No one man can tell another what to produce, how much to produce, or how to distribute his wealth. A man can do whatever he likes as long as it does not interfere with another's right to produce. He will most likely choose endeavors that are of value to his neighbors as well as himself. This allows him to concentrate on what he is best at, and trade his own productive efforts for those of others.

He participates in a government that mediates any dispute of this issue, which is one of the proper functions of government. He is safe from encroachment of his rights at home and from abroad, which are the other two functions of a proper government.

I have described what I consider to be a "natural" form of government. A truly moral government does not exist in the real world, as yet.

When I speak of government as a moral idea, I am referring to it as a cooperative effort of the many to protect the rights of the productive individual. This power of government, however, has been historically misused against the very individuals it was designed to defend: The producers. Perhaps this segment of society has been too busy producing—dealing with reality—to immerse itself in the kind of immorality and irrationality that a diseased government, whether it be despotism, collectivism, or a corrupted democracy, can breed. They are actually attacked for a reason for which they never thought they would have to defend themselves. They are attacked because they are productive, because they are moral. All governments seem to eventually bite the hands that feed them.

The taxpayer is a victim who is threatened and bullied. Anyone that complains is called unpatriotic. Those who disobey the non-objective laws designed to fleece the producers of their earned property are fined or jailed.

Under this kind of pressure, the moral have done something that they never should have done: They have apologized for being moral. They defaulted on their own values. They condoned the legislation of unfair tax laws and regulations. They paid the bribes. Every now and then, they half rear in indignation. But it is too late. The guns are truly at their heads. The saddest thing is that the guns are of their own manufacture. No nonproductive individual could ever have amassed the materials, invented the design, and manufactured the end product. He can, however, pull the trigger.

This issue is poignantly illustrated in America by the rivalry between the so-called "liberals" and "conservatives." The liberals scream that the "needs" of some (the poor, elderly, sick, minorities, etc.) must be met by abridging the rights of others (usually businessmen—also a minority). The conservatives don't really believe this childishness but do not have the moral courage to fight it. They are content to give up a little bit at a time to maintain the image that they are "conserving" the status quo.

The liberals are at "fault," and the conservatives have "defaulted." Perhaps the conservatives are worse.

GOVERNMENT CORRUPTED

The fault of immorality is not hard to understand. It is the old story of trying to get something for nothing. Some people want a cause without an effect. They want the proverbial "free lunch." They want everybody to be well-fed, well-clothed, well-housed, self-fulfilled, and eligible for free medical care. Who doesn't? But they will give no rational thought as to how this will all be paid for. They are acting irrationally by insisting that this is possible.

They front their so-called "humanitarian" operations with those kind-hearted souls who, though genuinely concerned about their fellow human being, are fiscally naive. A system of transfer payment programs is created that rarely benefits anyone more than the bureaucrats and administrators who cream their share off the top.

And what of the individual who demands the freedom to be of benefit to those he wishes? He is told that he is inherently evil. He would never help another unless he was required, so all of his excess income is subject to taxation and confiscation. As we know, this system is backed by force.

The lust of the irrational for the unearned has reached epic proportions. They want more than is humanly possible. They want more wealth than there actually is. To accomplish this, they needed to "create" money out of thin air. As a result the moral individual has had his tangible method of exchange in the marketplace relegated to the realm of the irrational. Most governments on this earth have participated in the destruction of the true instrument of money.

MONEY

Money was one of the most brilliant inventions of all time. We know that production is life-giving and moral. We agree that it is the responsibility of the individual to produce for himself, his loved ones, and those others of his choice. We recognize the fact that we are not alone on this earth. Therefore, we should also recognize the potential value of cooperation. This cooperation necessarily takes the form of an exchange of a value for a value. Man's greatest contribution to the general well-being of everyone else on this planet is not charity, therefore, it is work.

There are many other producers of other talents, abilities, and energies to work for or work with. But how do you compare one man's production to that of another? It is a very private affair. The two involved must agree. For example, if a shoemaker wants eggs from a farmer, they must both decide how many eggs equal a pair of shoes. Obviously this gets complicated shortly after this easy analogy. What if the logger who cut the wood for the henhouse wants shoes but the shoemaker does not need logs? What if the rancher who raised the cows that supplied the leather for the shoes wants some eggs, but the farmer does not need a cow? Cooperation is already difficult, as one can see.

Moral people needed a way of turning productive efforts into a generic, inherently valuable commodity. They began to use money. Gold was good for this purpose for three reasons: First, it is obtained by work. Someone has to go through the real, productive effort required to accumulate it, whether by trade or by digging it out of the ground. Two, it is practical. Gold is durable and submits easily to assay. Three, it is an eloquent symbol of the life-giving, productive activity it was chosen to represent: It is beautiful.

People agreed on gold as a standard value. At first it was minted into coins to provide a ready measure for its worth and to make it portable. But the world of commerce became very complex, and substitutions or receipts, called currency, were printed, along with non-gold coins. The gold was stockpiled securely in central vaults, like Fort Knox in the United States. A morally responsible and cooperative exchange system was to remain in operation.

This practical solution to world commerce has been corrupted. Today our currency is not backed by money. Governments took and kept the true commodity of value, gold, and left us with only the substituted pieces of paper that they print at their seeming whim. Can you see that printing an extra zero on a piece of paper does not equal the work required of digging ten times as much gold out of the ground?

The non-realists, however, don't have time to dig gold out of the ground, or even wait for the productive to do so. Their "humanitarian" plans are too important. That is why they destroyed money and replaced it with inflatable currency.

Governments turned the task of producing this counterfeit money over to a separate institution known in America as the Federal Reserve. They reserve nothing but the right to loot the producers. They trade paper for productive activity. When you try to trade the paper back, you get exactly what it is worth: Nothing. It used to be that our currency represented, and was redeemable for gold or silver. That promise was replaced with the words "In God We Trust," which I consider to be blasphemous. God neither took the gold, nor is holding it in trust. Now our paper "money" represents nothing and means nothing.

The only reason that you can get something productive from another human being is because we are all caught up in a myth that the paper has worth. We "trust" that the other person will honor this otherwise useless piece of paper and he "trusts" that we will honor it back.

By fault and default, paper money is forced on us, and we are unable to reject it. Money, however, is a small, though sad, example of the illness that threatens us. The more complete story is that man has allowed his most moral political idea, government, to become his nemesis.

Americans have come closest to the reality of a government that protects individual rights, but we have never acknowledged the moral basis for our own system of life-giving production: The free market, backed by a rational monetary system. We have allowed both to be corrupted. Now there is no way that we can stand in front of the world with the confidence that our system is the best. Does America truly stand for freedom anymore?

The government, slyly, has shifted the blame for this tragedy from themselves to the productive.

The stereotype of the greedy businessman has been firmly ingrained in our minds. An often used example is that of the "robber barons" who exploited the early development of the railroad industry. What people seem to forget, however, is that most of the problems occurred after the railroads became government regulated. Privilege and pull replaced ability in securing railroad work. The market no longer determined routes, prices, and services. The government, based on solely political motivations, determined these essential items. As a result the railroad industry is dying a prolonged death.

It is interesting to note that the people of this country, again, sensed that something was wrong, but could not grasp the reason why. They were sold the falsehood that the free market was to blame for the railroad "tycoons" without any evidence to back it up, except the inherent evil and greed of "big business." The cure was "big government," an antidote that has been far worse than the supposed poison.

GOVERNMENT INTERVENTION

The view of the government (a particularly nonproductive entity—observe the national deficit) as our sole defense against the evil businessman is a myth. Somewhere the faculty of reason has to be suspended. This is no problem as long as there is enough productive activity occurring somewhere to feed everyone else as the charade is played out.

I don't want to create the impression that all businessmen are perfectly moral, perfectly productive benefactors of society. Some are certainly evil. Some may even start off morally justified and become exploitive later on. Some may merely be of mixed premises and unable to differentiate between what are and what are not ethical business practices. The point that I am making is, again, a philosophical one: We need the productive efforts of businessmen to provide those goods and services that sustain us and improve the quality of our lives. When market forces are permitted to operate naturally the moral and productive businesspeople will rise to meet that requirement. The market is also the cure for the greedy businessman. When a businessman cannot provide value for the money that he asks, he will be put out of business by some entrepreneur who can.

Government intervention in the market place is potentially very dangerous.The immoral businessman "buys" into the corruptible government. A symbiotic relationship is created: Two parasites feeding off each other. The greedy businessman feeds off of the special privileges and power of the government, while the government lives off of the unearned "profits" of its "business partners." Death of both is only a matter of time.

This issue of interference in business is only one example of the problem with government favor-granting. A politician is constantly confronted with the moral dilemma of whether to protect the individual or succumb to the pressure of special interest groups. He is constantly tempted to use his office to provide for some at the expense of others, in return for the almighty vote. A peculiarity of the nature of man's free will allows him to fool himself into believing that he can receive something for "nothing." The nature of politics allows him to receive it. The politician, who is paid to be the guardian of individual rights, is often goaded or pushed into becoming an arbitrary violator of those rights. He can use his potential power (backed by force) to commandeer economic power.

To stay in office, a politician needs votes. In order to get votes he must represent his constituency. What happens when the constituency is looking for a "free" lunch? The politician is tempted to supply it. Therefore it is the productive person who must be robbed (taxed) to provide it. I could write much more about the mechanics of our tax system, but it certainly is not within the scope of this book. Instead, let me provide a different kind of example of how the government enacts seemingly beneficial laws that actually penalize the productive.

Right now there is a bill before the Senate to raise the minimum wage. Those senators who are sponsoring this bill support an implicit and essentially statist platform which is that man is inherently selfish and it is the government's responsibility to care for the needy because the people of this country cannot be trusted to help their fellow man without coercion. They determine that the minimum wage is, say, 27% too low. Therefore, they set out to raise it in the "public interest."

What happens when the minimum wage law is put into effect? It is supposed to raise the living standard of the worker. What it creates is unemployment. Now the cost of labor is artificially too high for businessmen to afford, so they cut back production. A business, unlike the government, cannot operate indefinitely at a loss. A business cannot print its own money, either, or sell bonds on tax revenues that won't be collected for years. Hence, more people (usually those on the low end of the pay scale, the needy whom these statists are professing to help and who will vote for them) are unemployed while a few others get a hike in wages. The corporation is the fall guy because it laid off the workers; but the government is directly responsible.

The government strengthens its position as this all goes on. Some few people will benefit from the law and feel that they should be grateful. Business becomes more cutthroat, as the corporations find it even more difficult to turn a profit. Business seems evil compared with this benevolent government. With the support of certain politicians, labor unions sprout up to protect the worker during these hard times by promising them jobs and more money. The unions support themselves by collecting money, in the form of dues, taken from the productive who have managed to keep their jobs. And of course we have new people on welfare, needing support from the government.

Now the politicians have even more people looking for a free lunch. Hence, more votes are offered in exchange for special considerations. The cycle will perpetuate itself until there are too few people left producing to feed the ones who aren't.

Obviously, the minimum wage law, like all laws that provide for the redistribution of wealth or the redirection of the free market as a response to the "need" of one group at the expense of another, must be repealed. This, however, solves only a piece of the problem.

Our political system must be refocused to protect the rights of all people equally, according to the natural law. Justice, not special favors, should drive our government. And our economic system must be freed from controls that interfere with the natural laws of supply and demand, providing value for value according to a standard all humans can agree upon.

So far I have presented a very rough analysis of some of the intellectual ills that afflict our government. There appears to be no place for a rational-thinking person when so many people of apparent good intentions have tampered with the inalienable rights of basically moral, productive citizens. The distinctions between right and wrong have been obliterated by our government. We have replaced natural laws with man-made ones that do not correspond to our true nature. There is a growing sense that *something* is not right with the way things are, but no sense of what can be done about it.

A SOLUTION FROM THE EAST?

In America, we have had several decades in which people have sensed that they are being attacked and should do something about it. They feel, and rightly so, that the attack may ultimately come in a very physical form. Some have looked at self-defense skills as a possible recourse. Here comes another disappointment: Oriental martial arts.

With a typical inability to approach things from a philosophical and moral base, Horace Greely's young man has gone east. He has adopted, indiscriminately, not only the martial forms, but the thought, religious tenets, food, and so on, of the Asian cultures which have mass-marketed their styles of martial arts. What we have forgotten to do is our homework. If we had, we would find that the legacies of some of these countries are rife with incidences of repression, xenophobia, and downright misery. Their sorry record on the subject of individual human rights has been documented far too openly to be ignored.

I, personally, have never understood why it has become fashionable, for example, to admire the samurai, a character who seemingly longed to die at the whim of a tyrant. Or why it is considered chic to practice Eastern religions, totally out of any cultural context, that emphasize negation of self? Suddenly it is cosmopolitan to enjoy foods that are but an updated version of a starvation diet because they are more "natural."

Obviously there are good aspects to every culture, that is

not the point. The point is: Is it rationally valid, when the methods of your culture don't seem to work, to compromise and borrow different aspects of different cultures according to the expediency of the moment?

The answer is no. But that seems to be our form of remedy for the nagging fears that beset us. Then we cannot understand why a syncretistic combination of such diverse disciplines as, for example, est, macrobiotics, Tae Kwon Do, numerology, rolfing, and racquetball all result in nothing but a kind of "holistic" mishmash. It is unintelligible because it is a system without a premise. It is doing without thinking. It becomes an exhausting treadmill that keeps the devotee in an irrational blur of methods that should have helped, but don't and methods that should help, but won't.

Of course there is nothing wrong with these activities if they are used as tools for a rationally definable goal. Alone, they do nothing substantial. At best, they can only provide symptomatic relief. But the problem usually just manifests itself elsewhere.

The crux of the situation is this: It is not enough to "sense" that something is wrong and then "feel" around for the right solution. Rather, one must rationally confront the root of the problem. To do that, one must hold a presituational value system which provides guidelines for interpretation of specific issues.

A RATIONAL APPROACH TO THE MYSTICAL EASTERN PHILOSOPHIES

It is a peculiarity of human nature that we cannot accept the fact that there are things that we don't know yet. Science has continuously proven that things we didn't know (such as where babies come from) are possible to understand, and that things we didn't think we could do (such as walk on the moon) are possible to do. Sometimes the answers are not what we expect, but that should be easy to accept when the proof is presented.

But, it is not science itself that provides the answers. Knowledge is the result of a rational brain functioning according to a rational epistemology. A rational, logical (Aristotelian) scientific *method* may allow us to derive those answers from experience. The method is accessible to anyone who will use his rational faculties. To learn what we do not know, however, requires a confidence (not to be confused with "faith") that we live in a world that is ruled by natural laws, and is, therefore, real, and potentially knowable. If you believe that the universe is unreal, chaotic, unpredictable, and unknowable, there is no reason to read further. In fact, there is no reason to ever use your brain again, because your brain, in such a world, is totally impotent.

Most people, on some level, reject the notion, inherent in many Eastern religions, that this world is some temporary purgatory of inexplicable torment. There is no evidence that we need to earn a better "next" life by suffering in this one. In fact, the opposite seems to be true. Observe the lengths that people will go to to make themselves happy, and how their attempts become thwarted or perverted only when they are interfered with or when their ends or means are not life-giving.

Nevertheless, people sense that there may be some real world value in the seemingly mystical Eastern philosophies. Some have tried to adopt aspects of these philosophies into their own lifestyles, or even embraced these religions in their entirety. I, myself, have found great value in the action-oriented philosophy of ninpo. That is why I believe it is important for me to explain the reasons why I believe the premises for most Oriental religions are at odds with my philosophy.

First, the dominant culture in Asia, namely, that of China, is probably the oldest civilization on earth. Therefore, they first possessed the tools of meaningful accomplishment (i.e., the means of obtaining wealth). It did not take long for the less-civilized hordes, such as the Mongols, to recognize that China was able to supply their lust for something for nothing.

Imagine the average farmers, merchants, artists, or scholars in primitive China. Struggling and slowly winning their fight to achieve their full human potential, they were making the first great strides to separate man from the beasts.

Over the horizon, time and time again, came the ravaging human animals to destroy what they were building. Each lull in the plundering was filled with repressive "government" activity, including the drafting and financing of government armies. It made no sense. It was illogical. Therefore, they (wrongly) assumed that life, itself, was illogical. To divorce themselves from the pain, they withdrew from the perpetual (it seemed) brutality of their physical existence into the recesses of their minds. I think it was quite understandable, under the circumstances. Why rely on reason when it is so easily overcome by unthinking beasts? How does one reason with the likes of Ghengis Khan?

The character of the Oriental mind is still plagued with this memory of irrationality. To their benefit, however, they have survived in spite of this. China is the most populous country on the planet. Torn between the temptation to view existence as an irrational illusion, and the rational need to live in the real world, life in the real world won out.

Of course, it *had* to. Life is man's most important value.

The benefits of their introspection include psychological methods, such as meditation, that may be of great value to man. Their esthetic sense is beautifully apparent in their unique style of art.

Still, in a world where the primacy of inalienable rights has been validated, the premise of the Oriental view of life as an evil to be endured is ultimately quite harmful.

The warrior knows that life is essentially good and that what destroys life is essentially evil. The philosophical error in most Oriental doctrines is that the occurrence of misery (always brought about by irrationality, if the protagonist is man, or ignorance, if the protagonist is nature) is given as proof that life is inherently miserable. Once that false premise is accepted, the view of life as an illusion is a simple next step. Life is not always easy, but that does not mean that life is a punishment. Life is not always understandable, but that doesn't mean that the world is a totally unfathomable chaos.

Although ninpo was deeply influenced by many Oriental philosophies, when you speak to the practitioners you do not get the impression that this is their sense of life. Takamatsu sensei once said:

> . . . know that being happy is the most satisfying of life's feelings . . . Knowing that disease and disaster are natural parts of life is the key to overcoming adversity with a calm and happy spirit. Happiness is waiting there in front of you. Only you can choose whether or not to experience it.

How can one not sense the ninja's true love of life after reading this beautiful statement?

Ninpo bids us to embrace life and overlook its trials and tribulations. I hope that I have not painted a too unhappy picture for you, the aspirant warrior. The bad news is that we are living in a world which is largely pragmatic. People do what seems to work without relying on an underlying, presituational value system. Relativism has virtually become a social institution. Our government, schools, and religions have been corrupted to the point where their main purpose seems to be to perpetuate themselves rather than to evolve so that they may provide a continuing atmosphere of freedom and support for the productive individual. The unproductive among us seem to have free reign to beg or steal whatever they want, while the moral, productive people are doomed to work as their indentured servants.

But there is good news, too! A man cannot live without a brain. The non-realists who seem to hold the upper hand in life are not using their own brains, so they must be using ours. We, then, hold the power to defeat them. If we refuse to work for them, they will sink into the void of their own impotency. When they try to turn on us we will have the intellectual, emotional, and physical skills to defend ourselves, our loved ones, and anyone else we decide to defend. We *have* that power. We just need to learn how to use it. I believe that there are three main ingredients to living that good life as a warrior. They correspond to the three sections of this book.

Of course there are many overlaps when one tries to break down the basic methodology of the warrior into separate components. Many of the examples that were used in this first section will be employed again in the second section to make illustrations of a different nature. In the third section we will speak of physical fundamentals that allow the warrior to function in any physically combative situation. Keep in mind that there is no way that I will be able to illustrate every possible fight scenario and every possible response, anymore than I could present every intellectual and moral issue in this complex world. Most of them you will have to resolve for yourself. But the skills to resolve these issues, one by one, can be developed by the practitioner.

Although it may take years of studying the fundamentals to do it, I believe that a person is capable of negotiating a way through any problem to attain a full and satisfying existence.

I hope that the "fundamental" examples that I have employed in this first section will allow the reader eventually to develop the skills to defend himself against all sorts of intellectual and philosophical attacks, as well as the rare physical confrontation.

Remember that your logical brain is your best weapon. Do not be demoralized by the state of philosophy in our current time. The attackers of man's mind are ultimately impotent. They cannot live to spout their irrational life philosophies unless you feed them.

Regardless of their numbers, unthinking "beasts," human or not, can be overcome by one man using his brain. How easy would it be for a thinking man to overcome someone who did not have the facility of rational thought? Very easy. Then is it not the same with many? If the rational brain is seen as the critical factor, and the factor is zero, doesn't zero times anything still equal zero? Takamatsu sensei once said that a single ninja could defeat a thousand men or ten thousand men. This kind of thinking is what I believe he meant.

When dealing with the real world, it is vital to be capable of rational thought. There are many ways to develop clearer thinking and a sharper intellect. I would suggest the study of Aristotle's epistemology as a start. Remember to think in terms of premises—those presituational fundamentals that can steer you to a spontaneous understanding of the solution to a situational intellectual problem.

Many times it is the mere confusion that comes from persuasive, but conflicting data that makes clear understanding difficult.

TRAINING THE MIND

I would like to end this section of the book with an intellectually oriented meditation exercise that may help you attain a clear thinking process, free from situational distractions.

The subjects of meditation and visualization are important ones which I will reintroduce in a slightly different manner in the second section of this book. With practice, these exercises allow the practitioner to begin to channel inner energies for physical, intellectual, and spiritual growth. Because the methods are tri-fold processes, involving the physical, mental, and spiritual self, I believe that it might have been just as appropriate to broach this discussion in any of the three sections of this book.

It is important to remember, also, that ninpo taijutsu is a philosophy of action. Although I will introduce some exercises as if they are to be performed in a certain way, i.e. seated, the object of this type of training, for the warrior, is the development of a clear heart and a powerful intention in action. I do not, therefore, specifically recommend that one assumes the traditional lotus or seated position when meditating. Rather, I believe that this process is worthwhile, and may be even more appropriate for the warrior, when it is practiced with movement. Quiet reflection or even specific mental or spiritual training can be practiced while walking in a natural surrounding, swimming in the ocean or in lakes and rivers, or as part of taijutsu or junan taiso exercises. Certainly the list of possibilities is endless.

It is a mistake to think of meditation as only a mental or spiritual exercise. The entire process is very physical as well. For example, even the most simple of meditation techniques requires an appropriate physical setting and can yield results that manifest themselves tangibly in the real world. For that matter, have you ever sat in the lotus position for an hour? It can be an extremely physical test of endurance.

Much of the growth, however, manifests itself in the form of increased intellectual maturity.

When you initially attempt the first meditation exercise you will find that the mind can be a difficult servant to train. Don't be too hard on yourself, however, strive for happiness and understanding in life, not mere facility in mind games. It is only after a foundation of rational thought is obtained that you will safely be free to explore the spiritual world without the danger of becoming lost in an illusion.

It is the creative ability of a human being to see the bridge between dreams and reality that makes happiness possible. It is quite important to consider the possibility that your way to enlightenment has never been trod in the exact same way before. For each of us, the "setting," or the circumstances of our life, are different. Due to our free will, they can proceed in an infinite number of ways. Which way is the best? This is important, because, when the stage is properly set, man can tap into a world where everything that can be known is known. For the engineer, it may be the way a wing must be shaped for it to be able to lift a vehicle into flight. For the scientist, it may be the cure for cancer. For the warrior it may be the way to save life when there seems to be no hope.

The ninja had the reputation of being able to accomplish amazing feats by the power of intention. Terms usually associated with ninjutsu, such as mikkyo, kuji-kiri, and kuji-in, sound exotic, mysterious, and exciting. Many people have a longing to toy with such arcane abilities as mind manipulation, invisibility, and transmutation. My suggestion is that a sincere warrior aspirant should seek, instead, to understand normal meditation and visualization techniques first. If these methods seem to offer a down-to-earth, usable value, one can always practice turning into a crow or hypnotizing attackers in a sword fight a little later. For those of us who need relaxation or resolve in any discipline, meditation can help.

Let me take some of the mystical mumbo jumbo out of *ninpo meiso* (meditation techniques). No matter how long you practice, you will never achieve results that are anything but normal and natural. Forget the magic, that is the realm of the non-realists.

I will present two basic approaches to meiso, one in this section, one in the next. The first is a simple meditation technique. It allows the practitioner to clear unwanted noise out of his mind for purposes of relaxation or increased general awareness, or as preparation for another form of meditation, visualization, which we will explore in Part Two.

EXERCISE: MEDITATION

DISCUSSION

Have you ever talked to yourself? Of course. Maybe you call it thinking to yourself. Words, phrases, sentences are created in your mind. Who "says" them? You do, but not with your mouth. You can even "hear" what is "said," but not with your ears.

Many psychologists, philosophers, and spiritualists have tried to explain this phenomenon. I cannot. But I am not sure that it is necessary to understand it to use it. There seems to be a definite correlation between the control of the "inner self" and control of one's mental and spiritual well-being. Often we observe that people who are mentally (or perhaps spiritually) ill hear "voices" in their heads that they cannot control. Sometimes these voices bid them to do terribly evil things, such as kill themselves or others. For some, the voices merely confuse and distract the afflicted person to the point where they cannot carry on a normal life. To function, they require drugs, therapy or both.

For most of us however, the symptoms of an occasionally out-of-control inner self are merely annoying. We speak of the inability to concentrate, or "noise" in our head. Occasionally we find it difficult at times to have the "outer self" and "inner self" work together toward the same goal. Outwardly we are involved in one endeavor, while our mind races off on another distracting tangent. This is not always necessarily bad. Very often the inner self can find a solution to a problem when the outer self seems to bog down. Perhaps the opposite is true in other instances. For example, the answer to a particular problem requires logical thinking rather than a random "brainstorming."

Your goal when meditating, therefore, is not necessarily to force your outer self to dominate the inner self, or vice versa, but to allow them, when necessary, to work together more efficiently. The outcome of this cooperation may take the form of a more incisive intellectual awareness or a more active and inspired creativity.

In any event, this first exercise will allow you to begin to clear away the "noise" in your head and improve your powers of concentration and creativity. You may develop a better cooperation between the inner self and the outer self.

The following exercise is adapted from one frequently used by Stephen Hayes in his seminar work. I think it is valuable for the new practitioner as well as for those who are already experienced in meditation techniques.

There are several stages to the exercise. It is not possible or necessary to master all of them the first time that you try them. It may take months or years. Certainly, you should consider working on one stage until you are perfectly comfortable with it before proceeding to the next.

EXERCISE

Stage one. Select a place where it is unlikely that you will be disturbed for 20 minutes or so. The place should be quiet and dim. Occasionally you may want to have some kind of background sounds. I live near the ocean and enjoy listening to the rhythm of breaking waves as a background for meditation. There is also some indication that baroque classical music, such as Handel, Vivaldi, or Bach at largo time (40–60 beats per minute) can be an extremely good background sound.

Sit quietly with your knees open and your feet tucked in. You can sit on the ground, the floor, a pillow, or chair, but the important thing is that your back is naturally erect without being stiff and that you feel balanced. Rest your hands in your lap. Touch the tip of your tongue to your top teeth at the place where they disappear into the gums. Decline your head slightly so that your eyes are looking at the ground about three feet in front of you. You can shield your eyes with your eyelids and lashes, or close them completely.

Take a deep breath and hold it. Movement in your chest should be barely discernable, but beneath that you may notice that you can see your belly (actually your diaphragm) inflate slightly.

Allow your body—head, neck, shoulders, arms, wrists, hands, fingers, chest, torso, hips, legs, feet, toes—to relax around that big ball of air. Relaxation is the entire purpose of this preparation.

When you are totally relaxed, let the air out slowly. Observe how your body wants to remain in what should now be a totally relaxed, naturally erect position.

Perform this breathing exercise nine times.

Stage two. Become aware of your breathing. Air comes in through your nostrils, into the lungs, and inflates the diaphragm. Imagine the cleansing oxygen molecules being extracted by the lungs and fed into the blood stream. Imagine the worn and weary carbon dioxide molecules coming back from the "front-lines" and returning back up and out through the nostrils, and back into the atmosphere.

Perform this exercise nine times.

If you have never done an exercise of this nature before, you may find that you are becoming uncomfortable with the position you are in. This is natural. You may want to take two or three deep breaths a la stage one and conclude your training until the next day. As was mentioned before, this training can be extremely taxing physically. If your body cannot cooperate, wait for it. There is no sense in agonizing over an exercise that is designed to make you feel good. This goes for all of the other stages. Train only as long as the benefit outweighs the discomfort. You cannot rush this kind of training.

Stage three. Become aware of the noises in your head. You may be thinking about other sections of this book. You may be thinking about external noises, such as a plane going overhead. You may be thinking about a discomfort that you feel in your knee or back. You may even be thinking about something totally unrelated, such as your shopping list or responsibilities at work. Be aware of them all. You will find that you may be verbalizing the thoughts in your inner self. Leave all of the thoughts right there but try not to verbalize them.

For example, you hear a siren in the distance. Don't say

to yourself: "Oh, that's a siren, I wonder what it means?" This common, everyday response is unnecessary. Merely acknowledge the sound, and continue with the exercise.

The position that you are in requires some training, also. Initially, you may find that the most difficult part of the exercise is physical. There will be aches and itches and feet that want to fall asleep. When they come, simply observe them, acknowledge them and put them aside. Don't verbalize them. Let them, and all other random thoughts, rest silently in your inner self. To help yourself perform this endeavor, give your inner self a task. For example, you can repeat to yourself the words: "I am relaxed" over and over. Another option, a bit more difficult, is to count a specific number of breaths. For example, count nine breaths and then nine again and again. When you feel a tangential thought trying to break into your rhythm, count "louder" and try not to let the other thought become verbalized. If you are counting and you find your attention drifting, you may realize that you are counting "ten, eleven, twelve . . . etc." Stop yourself, and go back to counting one through nine, one through nine.

These tasks are merely a way of substituting the random verbiage in your mind with verbiage of your own choosing. It serves to allow you to train and discipline your inner self. With long practice, you may be able to discard the use of the task that you have chosen to clear your mind of the "noise."

Stage four. When you are extremely relaxed and clear, you will find a gentle euphoria creep over you. The black of the back of your eyelids becomes darker, and you may feel a very clean and sweet saliva fill your mouth. Continue to wipe the "slate" cleaner and cleaner, swallow gently and go deeper and deeper into this relaxed state.

Start with ten minutes in this state. Stretch it to twenty in a week or two. After that, you can maintain it as long as you wish and are able to.

Stage five. When you are ready to come back to your normal level of activity, feel yourself withdraw from the warmth and darkness. Take two or three deep breaths and relax, as you did in stage one. With each breath, let yourself come closer and closer to the awareness of your immediate surroundings: the noises, the feelings, the smells. Gradually open your eyes and rest quietly for a few moments.

You may now stretch and stand up slowly.

I believe that there is great benefit in developing cooperation between the inner self and the outer self. Initially, you will gain an ability to calm and relax yourself when you need to. You may soon find that this feeling of serenity will last throughout the day.

Your ability to concentrate will improve. You are training your inner self, and, the resulting clarity of mind will aid in problem-solving and learning. The inner self will become more creative, as it learns to differentiate between inspirational and confused thinking. Remember, the point of this exercise is not to enslave the inner self to the outer self, or vice versa, but to allow them to join their unique forces as one tool of your intention. Your goal is to develop a rational thinking process that won't desert you, even when confronted with seemingly persuasive arguments that violate the natural laws.

Intellectual attacks, however, are only the first level at which you may be assaulted. Therefore, the use of your intellect to protect yourself against those who would harm you is merely one level of the ninja's power. When it becomes apparent that you are not susceptible to intellectual attacks, your spirit may be attacked. The next section deals with making your spirit strong.

PART II

THE SPIRIT OF THE WARRIOR

The most profound value of the warrior is human life. The only question is: Whose life? Ah, the answer to that is not as easy as it seems. I believe that this issue is at the core of what makes a philosophy good or bad, and therefore, what becomes manifested as good or evil in the world.

Primarily, we are led to believe that all men are equal. Equal how? Obviously we are not equal in wealth, intelligence, skin tone, productivity, or any other such qualification. Then, you may ask, how are we equal? Again, it is important for the warrior to understand a man's equality vis a vis other men. The philosopher must know why.

To illustrate this point I would like to relate a story that was told to me by Dr. Robert L. Humphrey. His credentials are too numerous and various to list here, but suffice it to say that he is a warrior and a philosopher of great standing in my eyes. This true story about a village in Turkey changed my view of life:

On weekends, the Americans would form parties to hunt the wild boar which were destructive to the farmers' crops. As the hunting party would go into the village, the more curious of the local farmers would meet the American hunters and crowd around the trucks.

The sight of those peasants in the poorer villages was often depressing. Many of the villages were only a few miles off the highway which connected the larger cities, but they were hundreds of years behind the cities in economic and cultural development. When the rains came, the mud spread like wall-to-wall carpeting in the streets throughout the villages.

Upon arrival of a truck full of American hunters, the villagers always gathered in the dust or mud to welcome them. Almost inevitably the sight of such a group of ragged, destitute villagers drew comments from the Americans such as, 'Look at them; they are like animals. What do they have to live for? They might just as well be dead.' What could anyone say about such comments? They seemed true enough.

Then one day in response to the familiar comments, an old sergeant drawled out his answer between spits of tobacco juice. He said, "You better believe they got something to live for. If you doubt it, get out there and try to kill one of them with your hunting knife. They'll fight you like no one you ever heard of. I've fought beside them in Korea, and I don't know either why they seem to value their lives so much. Maybe its those broad-beamed women in pantaloons, or maybe it's those dirty-faced kids, but whatever it is, they seem to value their lives just as much as we do ours . . ."

The lesson regarding the value which the humble village Turks placed on their lives does not stop with their tenacity in combat. Those same Turkish soldiers who fought in Korea also clung to life just as tena-

> ciously when the enemy was illness, despondency, and hopelessness in the North Korean prison camps.*

And this story contains the clue to understanding true human equality: *Every man is equal because his life and the lives of his loved ones are as important to him as yours are to you.*

For man to live according to this most important value, he requires personal freedom. When men are not free to live by this, their most basic value, the misery begins. Guaranteed.

Most of the national constitutions in the world today—even the Soviet Union's—seem to be designed to protect the lives of citizens. So why is there such inequality, human suffering, and lack of freedom? The answer is simple. It is a question of ownership. Who owns a man's life? The state? The church? Society? The Emperor? Or the man himself?

Obviously, I think that a man owns himself. Only a man can truly appreciate the value of his own life, and therefore choose what to do with it. Only an individual has the right to make decisions that most deeply affect himself. Most importantly, that includes the decision to sacrifice himself for others. Why is this necessary and moral? If we think of the primary human value, human life, as a dual value, it becomes a bit easier to understand the reason.

THE HEART OF THE WARRIOR

We have often heard that self-preservation is the primary law of nature. Is this completely true? If given the hypothetical choice, would a man save himself and allow the rest of civilization to go down the tubes? Well, probably not. A parent, faced by the choice of being killed or seeing his child killed might sacrifice himself. What about the soldier who leaps on a grenade to save his comrades? How

*From the book, *Paradigm Shift*, 1984, Robert L. Humphrey, pp. 30–31.

about young men who volunteer for military service without knowing why; except that they feel it is "right." It is clear that, without man's tendency to occasionally opt to sacrifice himself on behalf of others, our species would have become extinct a long time ago.

Most representative of this phenomenon is a mother's willingness to sacrifice for her child. Without this natural tendency on the part of the mother, the child could never survive infancy. Of course there are many other examples of personal sacrifice that preserved the group. The life value, therefore, is a dual one: Self-preservation balanced by the preservation of the group or species.

The question becomes: When a choice arises as to which side of the scale is preserved in a given situation, who makes the decision? In a collectivist society, it is the group. In a totalitarian country, it is the ruler. In a free nation, it is the individual. When I say "free nation" I am referring to a society in which there is protection of the individual's freedom to make life- or species-preserving decisions as long as those decisions do not violate another individual's same right.

There is much evidence that the species-preserving half of the life value is dominant in times of crises (war is the best example). Perhaps, this is why totalitarian and collectivist societies spring up during times of crises. Of course, they are usually just replacing each other. Collectivism is the usual cure for totalitarianism. Totalitarianism is necessary to provide a concentrated defense against collectivism. And so it goes.

The only alternative to this vicious cycle must be a free society. In a free society the *individual* makes a conscious decision, on a case-by-case basis, to pursue a self- or species-protective course. As with the soldier who protects his fellow warriors with his own life, the individual must be given the opportunity to make his best decision.

It is important to differentiate between a right and a privilege. A right (as in "inalienable" right) is an innate possession, such as the right to one's own life, liberty, and pursuit of happiness, as long as one's actions don't conflict with the rights of others. When there is a possibility of conflict, one's actions must be circumspect. For example, the operation of a motor vehicle may severely conflict with the rights of another if that operation is improper. Therefore, we are granted a license and a conditional permission to operate that vehicle, as long as we abide by certain rules. Driving, therefore, is a privilege, not a right. This privilege can be revoked if it can be demonstrated that we used that privilege to endanger others.

The acceptance of the responsibility of privileges allows men to live together in a free society. A man belongs to himself, so the rights of others cannot belong to him also.

For the Japanese ninja, this issue is probably illustrative of the greatest departure from the usual Japanese norm as eventually practiced by the samurai elite. The samurai belonged to anyone *but* himself—the emperor, the shogun, his liege lord, his father. The ninja was sworn to defend his community, his family, and himself, usually in that order. But the decision was his. He was taught to think as an individual, and made his best decision as to what actions he should take on a case by case basis. During wartime, of course, the ninja operative might relegate his personal desires to that of his commander. That is what all good soldiers do in strategic situations. But, the ninja had no loyalty to a ruler or ideology if it did not make rational sense.

Masaaki Hatsumi once told me that it is the job of the priests to "save" people, it is the function of the warrior to defend them. This is the way that the love of the warrior manifests itself: In the defense of the inalienable rights of self and others. Love, for a warrior, therefore, is a value system. It is based on the life value—the defense of self and others, according to an individual's choice. It is moral to preserve oneself. It may be moral to preserve others, *if you want to*. This defense is moral only when it is given and received freely.

The question becomes: Who is moral enough to make

such a decision? The answer is no one and everyone. No one owns another, therefore no one can make a self- or species-preservation decision for another. Everyone owns himself, and must make the decision—for better or worse—himself.

This is why such institutions as the military draft and tax-generated welfare systems are improper. The draft requires young men to give themselves, perhaps against their will, to a cause that they may or may not choose to face death over. The welfare system allows the "needs" of some to be used as a license to redistribute the earned income of others, thus decreasing the quality of the earners' lives. Volunteerism, in both circumstances, is the only way to make sure that the self- and species-preservation balance is rationally and morally maintained.

The ninja trains to become the most accomplished and productive person that he or she (yes, there are female ninja, they are known as *kunoichi*) can be. Realizing that individual action is the most loving way to maintain the dual-life value balance, the ninja strives to clear his heart of the temptations to violate this law of nature. The temptations are hard to avoid. After all, short range benefits seem to be available for those who do. These "benefits" can be the hollow satisfaction that comes from stolen wealth, or the illusion of a better world which is supposed to come about after the property rights of some individuals are violated for the welfare of the "less fortunate." The problem with this thinking is, that for every person above you on the economic scale who you feel could afford to contribute to the betterment of society, there is a multitude *below* who feel the same about you! It becomes a vicious system that knows no limits.

With all of our present day wealth, and all of the arguments that one level of society should be able to fleece another, certainly it may be a greater challenge, philosophically, to practice ninpo today than it was 500 years ago. The skill of the beggars has created an entrenched immorality which revives the myth of a heroic Robin Hood that steals from the rich and gives to the poor—the operative word being "steals." There should be no rationalization for stealing in a free society of producers. Since the industrial revolution, the activity of the producers has raised the standard of living to the point that the beggars can live comfortably, as they attack the very source of their comfort. This is the physical reality that we live in that is so different than that of the ninja 500 years ago.

Today a ninja seldom is called upon to scale castle walls or knock a samurai off a horse with a shuriken. Although the training is still very physical, the ultimate goal is to clarify one's values and "polish ones heart like a sword." Not surprisingly, the core of the matter is love.

SEEKING A BALANCE

Hopefully, few readers of this book will, in this lifetime, have to escape being murdered by a corrupt government to demonstrate that they love themselves. By the same token, let us hope that we will not have to prove our love for our fellow man by leaping on a grenade in a distant war or "police action." Let us, instead, take up the challenge of living a life of rational self interest balanced by reasonable concern for others. As warriors, we defend the rights of the individual to make each of the decisions that affect that balance without physical interference. Many governments have tried to take up that challenge in place of the individual and failed. Because we are all human and incomplete in total knowledge and wisdom, we make mistakes. *When an individual makes a mistake and defends himself or his community too aggressively, it is a relatively small matter. When a government does it in the "name of the people," a catastrophe occurs.*

MILITARY SERVICE

When a nation is attacked, it is proper for the citizens to take up arms to protect the nation, if they so choose. Conscription, on the other hand, is immoral. If a few individuals choose, correctly or incorrectly, to avoid military service in time of war, it is a small matter compared to the total usurpation of everyone's rights by the government. Incidentally, the dual-life value system operates in most rational people most of the time. If it occurs that mass numbers of people are avoiding military service in connection with a particular conflict, it may very well be that the situation is not truly one of necessary defense.

Once a person volunteers to perform military service he must be prepared to function in an imbalanced, and possibly, irrational atmosphere. The preservation of the nation, rather than the preservation of the individual, becomes the priority. Although the individual retains his cognitive capacities and value system, he voluntarily subordinates it to superiors who have been trained (we would hope) to utilize their soldiers judiciously.

In the Iran–Contra affair, for instance, an intelligent, rational Marine Lieutenant Colonel subordinated himself to his superiors. By his testimony it was obvious that he was not an unthinking automaton. Our leaders must take care not to misuse such men.

WELFARE OF OTHERS

When it is not a matter of life or death, what is the rational way to help others? It is clear that a feeling person cares about the welfare of those less fortunate than he. But what can, or should, he do? In many ways he is constricted by the policies of his government. When 10%, 20%, or 30% of a person's income is taxed away it becomes difficult to come up with the extra cash to be expansively philanthropic. And why should he be so? After all, doesn't the majority of his tax dollars go to help those less fortunate than him anyway? The whole matter is very confusing, to say the least.

On the other hand, when the government's main responsibility is to protect every individual's rights, rather than to ensure some groups' income, institutional poverty becomes unlikely. A sad example is the welfare system.

To ensure the income of certain poor Americans (and their attendant multitude of social workers and bureaucrats), the middle class taxpayer has become an indentured servant to the welfare system. Certainly, there is the perception that the welfare recipients must rely on the government to live.

The point that I am making is that, in a way, we still have a kind of "slavery" in this country. The only difference is that the present system is even less productive than the old one. At least the old Southern "aristocrats" produced something of worth.

Of course there are people who are truly in need, such as orphans, the mentally ill, or elderly who have no family, and others. I wonder how many people, if relieved of the burden of having to fund a gluttonous, tax-supported social services system, would hesitate to expend a little of their leftover time, money, and energy to relieve a neighbor who was truly suffering helplessly? I tend to think that most people would. Furthermore, I personally resent the inference that, without the threat of imprisonment (which is what happens to tax dissenters), I could not be depended upon to help another human being who truly required my help.

Clearly, most, if not all social service programs should be voluntary. Comingling of social security funds with other unconnected causes should be illegal.

WORK AND CHARITY

Let's consider a hypothetical situation: You receive a bonus at work of $2000. Which is more moral: Should you give it to the starving children's fund or take a vacation to Acapulco? Let's see what might happen to the money in each circumstance.

First, you give it to charity. It must pass through the

hands of certain administrators (some extracting a salary, some not). There are accounting, reporting, and transportation costs connected with transferring this money (by this time it may or may not be in the form of rice, milk, or gruel) to the impoverished area. None of these transactions produces an added value. On the contrary, all of them diminishes the size of your gift. To make matters worse, in most cases, the control is eventually passed to the usually inept or corrupt government of that country. There, a portion of the food is either left to rot or it is appropriated to feed the coffers of the rulers. Only a fraction gets to where it really is needed.

Now, we take the Acapulco vacation. First, we go down to the travel agent to book the flight. The travel agent gets paid to perform that service for us, so now he can afford to feed himself and his family. A ticket is printed. Now the producers of ink and paper are paid and can afford to feed their families. We buy a new outfit for the trip. Clothiers from Hong Kong to Saks Fifth Avenue could be involved. We get on the plane. Boeing and Pan Am workers are now productively employed. We land in Acapulco. Now we contribute to the livelihoods of a multitude of taxi drivers, bartenders, restaurateurs, chambermaids, scuba instructors, and so on.

Perhaps most importantly, we receive the satisfaction of having a good reward for our hard work. It sure beats the hollow enjoyment of a tax deduction.

The most important point to understand by this admittedly cursory illustration, however, is not what we expended the money for, but what value the money had in each case. It is clear that there is more tangible benefit all around in the second case. Can you see that in the second case, the interaction between people was more direct, that value-for-value relationships were created?

Many people deplore the growth of the so-called service economy. I think that it is marvelous. We have grown so far from the purely survival-oriented material economy of the last several hundred years that people can actually do things for each other and receive something of equal worth in return. When the starving children are grown and no longer prevented from serving, i.e. working for and with others (and eventually, perhaps, having others work for them), such abject poverty will vanish.

In collectivist and totalitarian societies, the job market is strictly controlled to prevent this level of service-to-others personal freedom. In our country we have gone so far as to institutionalize poverty. After all, if the poor were to "disappear," what would happen to the welfare state and its attendant bureaucracy?

If you think that this kind of thing doesn't happen in the United States, consider the minimum wage laws. It is the low end (not the top end, as purported by its advocates) of the economic scale that is most harshly affected by this law. People are prevented from selling their products or services at market value. Since few can or will pay for services that are not worth the price, production is slowed, services are cut, and an entire segment of society must remain unemployed.

This scenario will require a rational monetary system and personal freedom to work. We also need a change in perspective that acknowledges the evidence that it may be work, not charity, that provides the best means for people to help each other. Of course, this may seem a bit heartless; but I submit that it is no more heartless than condemning an entire segment of society to a life on the dole. *That* way of life, so far, is the result of government-sponsored charity.

There is much evidence that, agriculturally, the world has the resources to feed itself. After all, we pay farmers *not* to produce in this country. Why is this permitted when people are starving to death?

Perhaps it is a matter of *distribution*, not production, that is the issue. The question becomes, therefore, which is the most moral way to distribute wealth? Is it through charity? Is it the forced redistribution of wealth of the productive to the non-productive via taxation? The results of several decades of such a system refute these possibilities. The answer is to allow people the freedom to earn the wealth they need to live. The service industry allows even those without capital to trade a valuable commodity—their own time and effort—for the things that they need. Who can say what long-term benefits would result in even such a small step as the repeal of minimum wage laws? At first the amount of money that can be earned by the unskilled will not be great, but the freedom that results when a man is allowed to work for himself rather than be forced to live on welfare, will allow for more personal growth and, eventually, more wealth.

The goal of government should not be to feed people. This makes them dependent. The goal should be to allow them to feed themselves. This makes them free.

THE FAMILY

The warrior protects his country by participating in military service when required. During peacetime he trains and performs productive work. Much closer to home, however, there is a constant cold (and sometimes hot) war to be fought. In order to understand this issue it is important to differentiate between different kinds of institutions: Those that we participate in on a personal basis, such as a marriage or a family, and those that are created as an adjunct of government (or business for that matter). I will use the one word in both contexts, but I do not believe that they are similar.

An assault on the most basic social institution, the family, appears to be occurring on three fronts: the physical, the intellectual, and the philosophical. First, inflation and taxation have required, in most cases, that both parents work out of the home. Inflation is caused by a government that prints (or creates electronically) more money than it has collected in taxes and other revenues. This dilutes the wealth created by productive means and makes our money less valuable. Tax money collected to support functions of the government other than national self-defense and criminal and civil defense for its citizens, has the same effect. As a result, neither the mother nor the father can remain home to rear their children. It is economically (i.e. physically) impossible for the family to remain a cohesive unit. The family hasn't *chosen* to live this way, as will be discussed below; it has been *forced* to. The answer to this problem must include the curbing of inflation and tax relief, as well as a return to a rational monetary policy and a reevaluation of the role of government. Not a simple task.

The second assault is an intellectual one. It is directly related to the physical one in that it addresses the matter of who is teaching children the skills of living if the parents cannot. That privilege has been indirectly usurped by the government via the school system and other institutions such as day care. No longer does a parent have any substantial input into the curriculum that a child is exposed to during his or her formative years. They don't have time to become involved, and the strong bureaucracies in place do not want the interference from the parents (or the teachers, either, for that matter). Again, it becomes a case of an institution deciding what is best for the individual. Small mistakes made by individuals, such as individual disruptive behavior, are replaced by the horrendous blunders perpetrated by the institution, such as the current drive toward conformity in our schools.

Of course good schools and day care can be vital to the overall well-being of the family. The parents benefit from a certain amount of time outside the home, away from the kids, involved in personal growth of a business, artistic, and social nature. The children, ideally, would spend part of their day in an environment that would nurture their interaction skills and further their education. A good example of a value-for-value relationship is voluntary cooperation between those people "outside" one's immediate family. There is plenty of occasion for cooperation at work, school, and play.

But our system, as it exists, does not fill the role of supporting the family unit. What happens when the separation, as in most cases, is not voluntary, but, rather, uncontrollable? Both the parents and child feel that the institution is at odds with the nuclear family. Perhaps, the parents don't have the money and time to ensure the quality of the school. The experience for the child may be a negative, rather than a positive one. Worse, perhaps, is when parental input is shunned. Children sense that the parents (and, ultimately, they themselves) are impotent in the face of the institution. This situation does not support the child in the development of a strong sense of the primacy of the individual.

For a total rectification of these problems, relief from the physical assault is required. In the meantime, however, school bureaucracies must be decentralized and schools should have to compete for students in order to remain open. When the parents and teachers are more involved, the scale is reduced so that errors in judgment and curriculum are small and isolated. A more service-oriented approach will most likely increase quality. Competition among schools will breed quality in service.

The third assault is philosophical (or ethical). It also begins with the school system. For people to learn things, they should be able to choose to do so. To know what to choose, the person must develop a value system. In other words, they must see value in learning something. Therefore, schools are involved in teaching children values as well as the three Rs. Is it any surprise that they teach institutional values rather than personal values?

The student, actually, gets more of a mixed message. "You can be an individual—as long as you don't rock the boat!" Lining up, dress codes and uniforms, severely standardized curriculum, lectures and memorization drills instead of interactive dialogue between student and teacher, pep rallies, and constant references to "school spirit," all give the child the same message: Conform! Since the child cannot forget that he is an individual rather than an institution, he cannot be but alienated by a value system that holds value for those of the ubiquitous "them," but dubious value for "me."

The institutional value system of conformity taught in schools, and later reinforced by government institutions of all kinds, must be balanced by a value system that nurtures rational self-interest. This kind of instruction would be the responsibility of the parents, if they understood it themselves (after all, they, too, are products of the institution to a certain extent) and had the time to teach it, as well as role model teachers.

The problem drifts over into the husband-wife relationship, of course. What is the purpose of marriage? Obviously it must address the dual-life value system. It provides succor, sexual enjoyment, and companionship for the individual partners; and it supports species-preservation by perpetuation and education of children. Of course there are marriages that are kept together "for the children," and childless marriages. These situations represent personal choices and may, when viewed on a case-by-case basis, be perfectly proper. However, what happens when one of the purposes of this dual-functioning marriage is usurped across the board by the institution? It throws off the dual-life balance and problems occur. Basically, the parents are relieved of a major portion of their participation in making sure that the next generation grows to be productive individuals and members of society.

Why stay married in that case? It seems just as appropriate to have a child and surrender him or her to the institution. One can go on, perhaps enjoying those quantity rather than quality relationships that correspond to the now-dominant, but artificially short-range, self-side motivations of the dual-life value equation.

Marriage is a rational relationship, in that it allows a couple to satisfy both their self- and species-preserving drives, if they want to. When this balance is upset by outside forces, there is less reason for the institution of marriage to exist. This is why divorce is so common in our time. Families must retain the privilege of rearing their own children as they see fit. Institutions, such as the school system, must remain a support function, not the other way around.

The species-preserving side of marriage manifests itself as the parents bring a child into the world and teach him or her the skills to be a productive human being. The self-preserving side manifests itself as romantic love. As with all relationships, physical, financial, intellectual, or spiritual, this must be one that is based on value. Both partners must provide value for the personal satisfaction that they expect to receive in the relationship. This value takes many forms. It can be the value of sexual union. It can be the value of companionship. It can be the psychic value associated with the appreciation of the uniqueness of another individual. It can be the sharing of burdens that are difficult to carry alone. It can be the value of joys too good to keep to oneself.

Although there are intangibles involved in any attraction between a male and female (including, presumably, the subconscious desire to mate—a species-preserving tendency), with time, it becomes apparent that, in healthy marriages, people stay together volitionally. That is, because they want to. That is, because there is value in it for them.

Not surprisingly, the same flawed philosophy that spawns large-scale catastrophes such as totalitarianism and collectivism can infiltrate a marriage. Gone is the option of individual choice. It is replaced by a lust for the proverbial "free lunch." We hear phrases like: "You would do that for me if you loved me," and "marriage is sacrifice," and ". . . that's what you're supposed to do when you're married."

Marriage, of all institutions, must be based on the freedom of the individual. Then the choices he or she makes are sincere and life-giving. Of course, a little bit of credit is allowed for both partners, but when the psychic debt is too high for one spouse or the other, resentment grows.

I often wonder why people, who profess to have done "everything" to save their marriage often get divorced anyway. The answer may be that everything that was done was not of value to the other person. The value-for-value relationship may have broken down.

I am reminded of a line in a movie in which a wife was berating her husband for being unresponsive to her needs and failing to do little things to make her happy as she does for him. Indignant, he says something to this effect: "Hey, I give you flowers." She looks him in the eye and says: "You give me flowers when you need to give them, not when I need to get them." Touché.

THE NINJA HEART

One might ask, how does one attain and maintain a rational regard for self and others when it is obvious that much of the world is not consistently moral? In some cases mistakes are made out of ignorance or a failure to think about one's motives and actions (default). Sometimes the mistakes are more calculated and evil (fault). The philosophy of the ninja allows the practitioner to protect himself from these immoral people.

This resistance to evil may take many forms. It may be superficiality. Since most people are never 100% right or 100% wrong in their attitudes there is always something to enjoy about a particular person and something to avoid. Begin casual relationships superficially. Take the cream off the top, if you will. If the relationship starts to develop beyond the surface, you may make a decision to set the example in situations that require moral decision making. Without starting an argument, you might mention that you do not agree with certain philosophically immoral or illogical statements that are made in your presence. Most people want to be moral and seek value in their relationships. These persons will be attracted to you if you present a consistent, logical philosophy. They may become good friends. The important thing is to be very subtle and patient. People own themselves. You do not necessarily have the privilege of showing them the "error of their ways."

Be sure to protect yourself from people who will attack you for being moral in order to temporarily relieve themselves of the hollowness of their own shallow existence, or to put themselves in a position to steal or beg from you.

A dangerous trap for the warrior philosopher is pragmatism. A pragmatist has discarded his value systems in favor of the things that "work." In an imperfect society, some things that are definitely immoral seem to work. Again, I am reminded of the military draft which seems to "work" in its goal to fill the ranks of the military forces, but violates the rights of all citizens who are conscripted against their will.

One of the famous historical attributes of a ninja was his ability to assume disguises. He could become another person in order to infiltrate an enemy stronghold. He would assume the dress, mannerisms, speech, attitudes, and habits that would allow him to be seen as acceptable in that foreign atmosphere. Of course he had to remain the person that he was, deep in his heart, so that he did not eventually become one of the enemy.

The warrior of today must be much more clever. He must maintain his presituational value system in a war that is much more insidious and subtle, a philosophical war, if you will. Usually the attacks are in the form of a seemingly innocuous phrase uttered by an acquaintance, or a "socially responsible" program being recommended by a local politician, or a small honorarium that is customarily given to this person or that. Sometimes, of course, there are obvious travesties of personal freedom that are now "law," such as the "voluntary" income tax laws, the violation of which would land you in jail. What does the warrior do?

Obviously he must live within the system. He cannot maintain his self-defense when he is behind bars. He cannot teach if he has alienated all who would listen. But most of all, he must not give up his value system because of the difficulties he will encounter in living a truly moral existence.

A moral existence is one based upon the dual-life value system of self and others. It is balanced by reason and exercised by the individual.

In this day and age, we are constantly admonished by the institution to be less selfish. Since we have a natural tendency to assume a self- and species-preserving balance in our lives, when the institution usurps too many of our choices, we respond selfishly. We tend to seek the balance. This should be a sign to the institution that it has overstepped its bounds. Instead we are faced by this absurd recommendation: Surrender *more* of your choices to the institution. Become *more* selfless.

One of the meanings of the word *ninja* is "one who perseveres." The life of a warrior is one of quiet perseverance. In each day there are many decisions that must be made. The challenge is to know when and how to make the moral decision without endangering yourself and your vocation. The warrior must take on the attributes of the reconnaissance soldier. He must correctly interpret the situation, and live as morally as he can, without subjecting himself to censure or punishment by non-objective laws. His mission is compromised if he is observed.

The most complete name of the art that I practice is ninpo taijutsu. Taijutsu is the ability to move freely and naturally in any circumstance, from the most benign to the most life-threatening. Ninpo is the admonition to move with stealth so as not to become a target for those who live by preying on others as they struggle to live their moral and productive lives.

VISUALIZATION: CREATING SUCCESS

The first section of this book dealt with the intellectual ammunition that a warrior requires to give sanction to his actions. The last part of this book is very physical, in that it is a guide for training the body to be a tool of attainment. The subject of this section is the *heart* of the ninja. The heart is the guiding force behind the life of the warrior, and as such, can be trained to prepare the warrior for spiritual growth. As we discussed in the first section, methods of meditation and creative visualization can be used to focus our intention and create successful accomplishment. As a followup exercise to the meditation exercise, I suggest that you explore the possible values of visualization.

I cannot tell you what should be in your heart or what manifestations your love of life will take in the real world, but visualization may help you forge your dreams into accomplishment.

In the meditation exercise our goal was to set the stage for the development of intellectual power. This next exercise, although a seemingly simple one, gives a clue as to how you may set the stage for the spiritual development that allows your intentions to become reality.

DISCUSSION

Have you ever felt strong, or successful? Of course you have. What if it is possible to take the energy and power that resulted from that success and keep it in your heart as a sort of power-pack, available for later use? You could recall the feeling you had at that particular time, become reinfused with that flush of success, and use the resulting energy to perform other difficult tasks. Such is the object of visualization.

Athletes use visualization all the time. The basketball player closes his eyes and "sees" the ball swish through the hoop over and over. The baseball player "sees" the bat hit the ball and sail over the fence. In martial arts, we observe the assailant's attack and "see" ourselves performing all the steps that make up an appropriate, life-protecting response.

The exercise I describe next is a bit more involved than merely closing one's eyes and seeing a completed act before it is begun. It contains several stages, as before. They are designed to more effectively prepare the practitioner for realistic success. Although there are many intangibles involved, it is still a rather down-to-earth exercise. You will need to concentrate and be clear in your goals to gain maximum benefit from the process.

EXERCISE

Stage one. Repeat the preparations that you performed for the previous meditation exercise. Allow yourself to become relaxed and your mind to become as clear as possible. This may take five or ten minutes. When you are ready, proceed to stage two. If you cannot attain the desired level of open awareness, you may want to consider postponing the exercise until another time.

Stage two. Prepare a short documentary for yourself. Call it "The Best of Me." Think back to the successes that you have experienced in your life. Accomplishments that made you feel good. Most of us have never won a Nobel Prize, but most of us can recall a minor victory of some kind: Making a good play in a team sport, receiving an "A" on a tough term paper, getting that long-awaited promotion at work, or falling in love together with someone. If you are a martial arts person, a major rank promotion or a genuine self-defense situation in which you performed well may be appropriate as a subject for this exercise.

Whatever the situation that you come up with, make sure that it is a great one, or at least one that can serve as inspiration for some other similar accomplishment that you would like to make.

Relive this occurrence in your mind. Revel in the glory—in your own modest way, of course. Imagine that the light of the sun is on you, warming you, bathing you in glory. See every detail of the situation. How did you feel? Great! You felt the power to accomplish anything. The light gets warmer, and you notice that the warmth is concentrated near your heart. You realize that the feeling, the glow is a generic thing. It is an entity that exists separately from the specific occurrence itself. Once you have the feeling, allow the glow from the sun to be mirrored in your own heart. Now it emanates from your heart. Hold the power in its pure, non-situational form in the center of your chest. Feel the power in you.

Stage three. Now that you have the power, imagine some other accomplishment that you would like to make happen. Be reasonable. All things have an appropriate timing. As important as it is to reach for the stars, it is just as important to know what you could realistically accomplish next.

Perhaps your goal is to succeed in school, or to improve your performance of a particular assignment at work, or even to quit smoking. Visualize all of the steps that it would take to accomplish your goal.

One of my goals was to go to Japan for training. I imagined the preparations I would make. I imagined myself in an airplane seat, going through customs in Japan, negotiating the train system, finding the inn that I would stay at, meeting with Hatsumi sensei, and finally, training. At first I didn't know how I would get to Japan, where I would get the money, how I would be accepted by the other Japanese training members, etc. But I visualized the successful accomplishment of all of my objectives. They occurred even better than I had hoped. The plane flight was great, the trains were manageable, and the inn was lovely. Within twenty minutes of my arrival, Hatsumi was visiting *me!*

Select a goal for yourself. Start small and work up to President of the United States if you want to.

Once you have selected the goal, be very specific about all the details. Allow yourself to actually "relive" the occurrence as if it had already happened. Become the accomplishment.

Stage four. When you have clearly visualized your successful action, return to the metaphor of the power pack in your heart. Once again, allow yourself to feel the generic power of your own intention. Gradually let the focus dissipate until you are left with only the warm and happy feeling of the sun on you. Take a few deep breaths, relax, open your eyes, stand up, and go take that first step toward accomplishing what you have visualized.

My suggestion is to think big, but understand that you must create success one little accomplishment at a time. Use each previous success as fuel for your next endeavor. Eventually, you may build up a momentum that will take you beyond anything you can imagine today, if you perform the intellectual work and the take the physical steps that your dreams map out for you.

Remember, it is easy to conjure up things that you would like to do "someday." It is the mature person who knows what needs to be done *next* to accomplish a particular goal.

Often aspiring warriors come to me and ask, "What do you think I should be working on?" I have to tell them that I don't know. In a lifetime there are so many things to learn. Even our own warrior tradition has many aspects to it that some may never explore. Each warrior must decide what *he* needs to know next. Maybe then he can ask the right questions and train in a specific way to accomplish his goals. No one can help you until you know what your goals are. Keep this in mind as you train. Force yourself to remain the master of your own destiny.

Of course, I have only introduced you to a few ways of using the meditation/visualization process. Clearly, you will become bored if you meditate on your only Little League homerun every day. I suggest that you think back upon the other concepts that you have read in this book so far. Do they sound right after concentrated reflection? What do you agree with? What would you say differently? Why? If you succeed in clarifying your personal value system as part of your work in this area, I would say that you have accomplished a lot!

Visualization is a powerful tool toward accomplishment. When the necessary knowledge and action is also part of the process, the combination is virtually unstoppable. In

Japanese this amalgam of thought, word, and deed is called *sanmitsu,* or the three-part secret.

Clearly, this concept is a very powerful one. Take care when you process your dreams through this method. They might just come true!

The warrior way of learning the final piece of the three-part secret, action, is contained in the next section.

PART III
THE BODY OF THE WARRIOR

The impact of the warrior on his world is, ultimately, physical. He is a man of action. His actions, in most instances, may seem similar to the everyday activities required of all men who live on this earth. At other times his actions may be those required for him to prevail in life-or-death combat. This aspect of combat ability seems to be overlooked by some. I have even read books that proclaim that the most direct approach to gaining a warrior's heart is through seated meditation. I couldn't disagree more.

A warrior must be proficient in real combat skills. Of course, there is more to warriorship than mere fighting. That is why there are two other sections to this book. But to leave out the third is impossible.

I see the life of a warrior as that of a quiet hero. I think that we are losing our concept of what it means to be a hero. I first learned about the Japanese ninja in a James Bond novel entitled *You Only Live Twice.* I loved all of the James Bond books. James Bond was my kind of hero—quiet, resourceful, intelligent, and irresistible to women.

He was a man of action. I have since come to realize that James Bond was a very one-dimensional, incomplete character; but that has not diminished his charm for me or for the millions who are still inspired by this accessible hero. It has always bothered me that the later movie versions portray him as a geeky punster, who wins by gimmick. It can be argued that the movies are entertaining, but the point remains that it is a crime to destroy our heroes so easily.

The third section of this book is not designed to make you a hero, but I hope that it inspires you to be a person that can act quietly and effectively in your defense and that of your loved ones. I consider this subject of physical action to be the most important part of this book, and the piece that is so unsatisfying, or totally absent, from all of the other philosophy and psychological "self-help" books that I have read.

As a review, in section one we pointed out that it is up to the individual to be moral. The warrior is the quintessential, moral individual. In section two we pointed out that each individual must be free to make self- or species-preserving choices for himself. The love of the warrior becomes manifested in his willingness to defend that right for himself and others of his choice.

This section presents the basic exercises for achieving clarity of physical movement. Sufficient physical training, combined with proper motivation and understanding, will result in correct action. The warrior reaches enlightenment, not through mystical revelation, nor solely from logic, but by correct movement.

The word *taijutsu* literally means "body techniques." When the techniques are used as an expression of the glory of the individual in defense of the personal freedom of self and others, these techniques become elevated to the level of an art. Romantic words aside, the physical part of the warrior's training is long and arduous. Hatsumi sensei, when he visited the United States in 1981, spoke no English. Yet he wanted to communicate, in our language, the essence of his training method. The phrase that he chose to learn for that purpose was a simple one. He said "keep going."

Years later I am still thinking about that phrase—and still going; no end in sight! There *is* no end to the training. You keep going until you die. Even then, I suppose, we will still be "going" in some other sense.

But there is another side to this motto: "Keep going." On one hand it prepares you for the long years of training ahead. Yet, on the other hand, it is a warning meant to protect you from the danger of training too hard. Since there is no end to the training, there is no need to rush to "finish." You must train with a pace that offers *both* maximum growth *and* maximum enjoyment.

The training should be fun. After all, it is a means of assuring personal freedom for yourself and others. This should be a happy thing. In 1986, when Hatsumi sensei returned to the United States, this time he used a new English phrase that reflected this attitude. He would demonstrate a technique, then, with a wave, bid us go practice it with the instructions: "Go play."

This third section, itself, is broken up into three parts: a short one on health and diet, one on walking and junan taiso, and one on the fundamentals of taijutsu. These are the three pillars of our physical training method.

HEALTH AND DIET

For your body to function in a critical situation, whether it be stress at work or a vicious knife fight, it certainly must be strong and resilient. Tension in both environments—eventually in one—immediately in the other, can kill you. You have to take care of your body, this goes without saying, but, what is the best strategy? Do you "whip it into shape" or do you pamper it? There are many ways of thinking. For the purposes of this book, it is more important for me to offer some suggestions than to provide a specific regimen.

Basically, I believe that a realistic and positive attitude, supported by a decent diet, is the foundation for the best results. There are, however, some topics that are important to touch upon. Remember that the subject of this book is warriorship, not general fitness and health. The lifestyle of the warrior is quite removed from that of the person who is just "trying to stay in shape." It is up to the reader to decide which suggestions can appropriately be incorporated into his training regimen, and which are unnecessary or counter-productive.

AEROBIC EXERCISE

Since taijutsu is not a specifically aerobic exercise, I think that some aerobic exercise should be part of the training day. For some, walking can provide that. We will discuss it in more detail in the next section. Others prefer a more strenuous workout. I enjoy ocean swimming and jogging in the woods. Swimming is an important personal skill for the warrior. Learn how to do it right from a qualified instructor, then practice and enjoy.

When running, be sure to take small steps and keep your knees slightly bent. Absorb the hardness of the ground in your muscles, not in your ankles, knees, and hips. Don't bounce, glide.

You can practice your relaxation or visualization techniques as you move, if you like, but watch where you're going.

WEIGHT TRAINING

I do not want to say too much about this subject. Whether or not you want to involve yourself in this kind of training is up to you. Remember, however, that I use the word taijutsu as a metaphor for freedom. Exercises that increase your power *and* freedom of movement may be more desirable than certain exercises that build strength, but end up restricting movement or creating tightness in the muscles.

SLEEP

Obviously, sleep is very important. The problem is that it is not always possible to maintain a perfect sleeping regimen. If your schedule allows it, you should alter your sleeping patterns occasionally. Learn to sleep in the day, or stay up all night. There may be times when, in a critical situation, you may be required to stay up for long periods with very little sleep. Allow yourself to experience what that feels like. If you have to do it for real, the shock to your system will at least be a familiar one. If you travel often, as I do, you may become afflicted by jet lag. Use this opportunity to allow your body to learn how to function in unfamiliar patterns.

LIVING ENVIRONMENT

It is clear that we do not live in a perfect world. At times we may experience bitter cold, stifling heat, air pollution, noise pollution, and so on. Some people feel that it is best to get away from all that: Go somewhere where the living is easy and the air is pure. This is nice for a change every once in a while, but few of us can live in such perfect surroundings all the time. It is, perhaps, more important to be adaptable to all environments.

A gradual, then sustained, introduction to the outdoors can make us rugged and less susceptible to certain illnesses and allergies. We become re-attuned to nature and adopt a relaxed, long-term view of life.

Periodic visits to the city can build a tolerance to pollution, noise, and congestion. It allows you to develop the minute-to-minute awareness that makes sharing the world with millions of other people so exciting and rewarding.

Understanding which techniques, what clothing, and what pace allows you to function in very hot weather and very cold weather is a useful exercise, as well. As a Marine Corps officer I performed duties in many different climates, occasionally as part of survival exercises. This training has come in very handy over the years. Training of this nature is important for the warrior. There are a number of good books and courses available. I am not convinced, however, that civilian-run paramilitary training is at all necessary.

Some of my friends have asked me if, as part of their training, they should go out to the mountains and live off the land for a time, like Miyamoto Musashi, who wrote *The Book of Five Rings*, did. Of course it is important to have access to healthy and beautiful places. Occasionally the body or spirit becomes weak and requires recuperation. When Takamatsu sensei was ill with beri-beri and afflicted with tapeworms after returning from China, he went into the mountains to cure himself.

But, Hatsumi sensei told me that a warrior cannot live like a flower in a greenhouse. The person who insists on living in only perfect conditions is not walking the warrior path. He is a man on vacation.

DIET

Diet is also a very personal subject. I can only present general guidelines that have worked for me. Please consult a doctor or nutritionist before radically altering your eating habits.

Certain foods produce certain effects in the body, so it is important to understand a little about this subject. There is also a movement toward consuming more "natural" foods in our society. We should remember that what is natural for one person may not be natural for another. For example, there is much talk about the benefit of eating "natural" macrobiotic foods. I enjoy eating well-prepared macrobiotic food, myself. When I am sick I tend to eat it a great deal, as it seems to have a recuperative effect on me.

Yet, can I say that these are "natural" foods for me? Being from Irish and German heritages, perhaps the most natural food for me would be sausage, sauerkraut, and boiled potatoes. Certainly my ancestors used these food groups as body fuels for centuries. Isn't my digestive system a product of this successful evolution?

Of course, these are merely speculations on my part. The world has definitely changed over the centuries, as well. For example, I certainly don't toil in the fields all day, so the amount of meat protein and fat that I require might be much less than is usually contained in a traditional Irish or German meal.

To develop my own diet I did several things. I tried to draw from food groups that were readily available. I, then, considered what foods were probably natural for me, according to my heritage, and included them. I consulted with a professional and learned which foods supported which activities the best. For example, I learned that proteins and natural sugars are good for supporting periods of high-level activity. Carbohydrates are best for supporting sustained activity, but tend to produce a calming effect. Carbohydrates are also easier to digest.

With this information I decided that I wanted to wake up hungry in the morning and give myself a high grain and protein breakfast with no meat. For lunch I would have my daily meat intake and a good salad. Snacks would be mostly fruit. Dinner would be a carbohydrate feast and salad again. I would try to walk before going to bed, especially if I had been drinking alcohol.

Again, however, it is important to emphasize that living in a "greenhouse" is impractical. Too much of a good thing makes you weak. Periodically you should consider eating whatever is at hand. Try new dishes from other cultures. Eat a Big Mac.

There is some evidence to indicate that persons who eat only vegetarian foods gradually lose their tolerance for meat. There may come a time, in a critical situation, when meat is the only food available. You must be able to live on it if it is the only alternative.

Another part of the training may be to live *only* on vegetables or grains for a while. Or fast—don't eat at all for a few days. The body must be able to function in either circumstance. Before fasting or severely altering your eating habits, consult with a professional, as I have done.

I also suggest that you try eating only a percentage of what you usually eat, and see if you can get away with eating a little less. You may be hungry for a few minutes, but then realize that you are really quite satisfied. Hatsumi sensei recommends eating until 60% full. You may also consider eating less, but eat more often during the day to make up for it. Small amounts of food are digested more efficiently than large ones. Don't starve yourself, yet it seems clear that the amount of physical activity that the average person performs per day in our society does not require as much food as we tend to prepare for ourselves. Find out for yourself the amount of food that you *require* to function comfortably.

Get in the habit of consciously smelling and tasting your food. Excessive amounts of seasoning, particularly salt, may be unhealthy. It also will dull the senses. An acute sense of smell and taste can protect you from ingesting spoiled or poisoned food and drink.

Heavy spices also make your body and breath smell. This may give away your presence if you are hiding from an enemy. The same goes for heavy cologne or perfume, by the way.

This subject of diet is an important one. The warrior must balance his requirement for healthy, common-sense dietary habits, with the need to be able to function on the less than adequate or desirable. General awareness of what is being ingested into your body, a developed sense of taste and smell, and a working knowledge of the results of eating certain kinds of foods are all part of warrior training.

Finally, it is not enough for you to copy the eating regimen that I have developed for myself. Without the general knowledge acquired by personal research and dietary experimentation, you cannot say that you have a complete understanding of different eating patterns on your own body. You may end up merely making yourself sick.

Take responsibility for every aspect of your training. Don't take my word for it.

WALKING AND JUNAN TAISO

Very often I am asked the question: "How do I train by myself, if necessary?" This is an important question. There are only a handful of ninpo taijutsu teachers outside of Japan, they are not conveniently located near everyone who wishes to study this method. Work schedules or family responsibilities may further complicate matters, making it impossible to travel for periodic coaching.

Intelligence gathering is a characteristic skill of the ninja. Hatsumi sensei, in his book *Hiden No Togakure Ryu Ninpo,* translated by Rumiko Urata Hayes, states:

> In order to develop in the fullness of what the ninja arts entail, it is necessary to study foreign languages, history, economic theory, religion, philosophy, psychology, chemistry, physics, geography, and cultural topics in addition to what might be thought of as the more obvious skills of weapon and hand-to-hand combat.

Obviously this aspect of one's solitary training can be accomplished in the library or through university extension courses.

Another aspect of the training includes attention to the health and diet regimen, as presented previously.

Perhaps the most important training or preparation for advanced taijutsu skills is proper walking methods and junan taiso. This training can be performed alone.

WALKING

I cannot overstress how important walking is to the warrior. Besides the usual getting from point A to point B, it must be remembered that we walk the same in mortal

WALKING

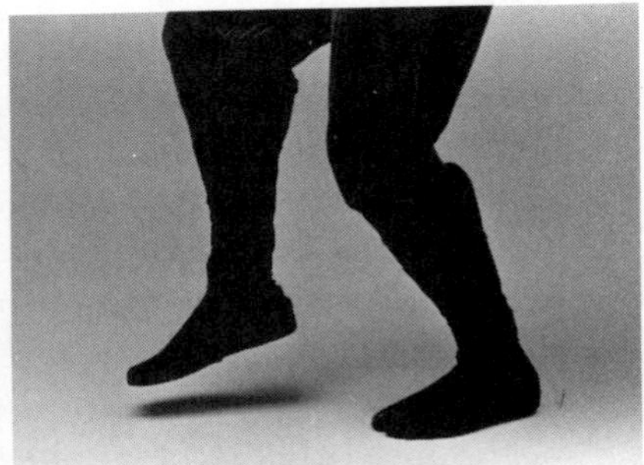

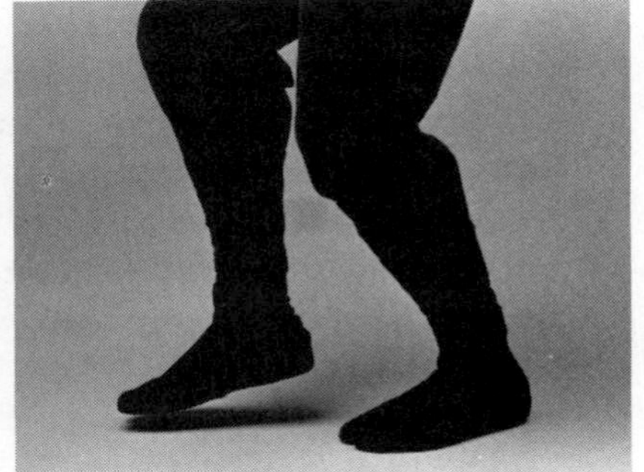

Walking motions, front view

combat as we do when we walk down the street. You cannot have two different strategies for walking. Proper walking is healthy and self-protective. It is also, literally, the strategy for moving in a fight. Learning how to walk properly takes a lifetime. You can never be too good at it.

I remember my days as a brash young Marine Corps officer. I walked, or "humped," as we called it, all over Okinawa, Camp Pendleton, and many other places, domestic and foreign. As if that wasn't enough, I used to run 5-7 miles in combat boots. I thought I was so tough.

I was so stupid.

By early 1982, I had pretty much crippled myself. I had crondomalasia in my left knee and tendinitis in both legs. I

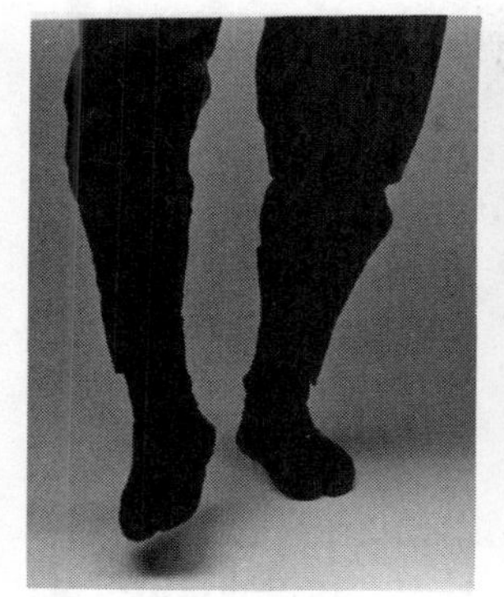 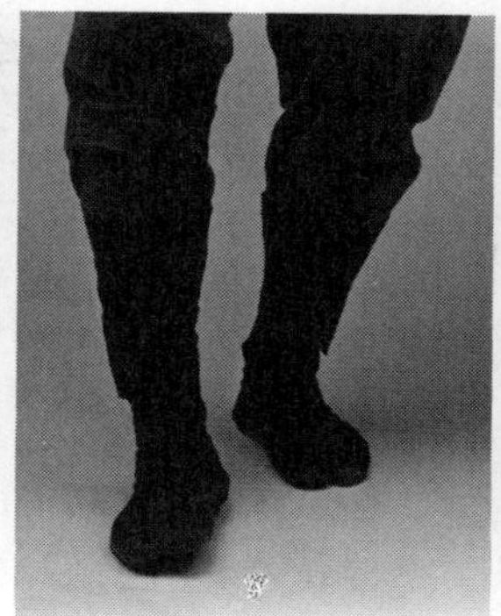 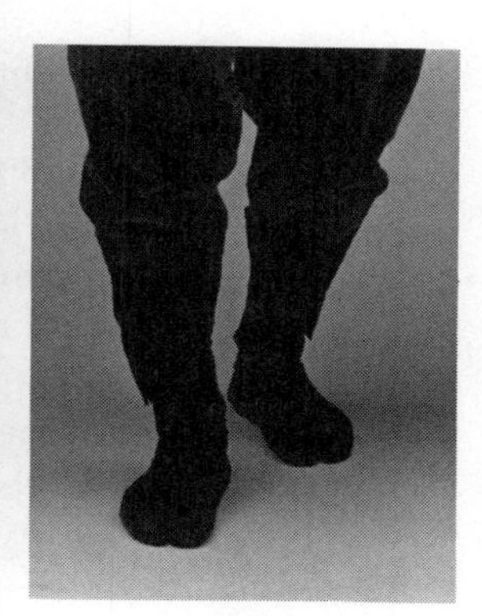

Walking motions, side view

was told by several doctors never to run again. The advice was not needed. I could barely walk, let alone run. Lucky for me, Hatsumi Sensei gave us all a lesson in walking that year. For the past six years I have worked on my walking every day! I can now get around quite well, and even run that 6–7 miles occasionally. I attribute it to learning how to walk naturally. I also use the same strategy for moving in a fight.

I have prepared some pictures to attempt to illustrate the basic ninja walking method. Pictures, however, do not adequately portray the flow of the movement.

Your body must be naturally straight. Feel as if there is a string connected to the top of your head that is holding you up, with your feet lightly on the ground. Breathe with an open chest. Arms and hands are held loosely at your sides, like a puppet's. Hips and ankles are loose and flexible. Knees are slightly bent. Toes are feeling the ground. Roll your foot gently from the outside to the inside, but don't over-exaggerate any of these movements.

Don't take big steps. I concentrate on taking the smallest steps I can without looking like a nut when I walk down the street. You will find that your balance will improve overnight if you shorten your steps.

When walking, try to walk as if you don't care if you fall down. Align your hips, knees, and ankles so as to allow gravity to propel you forward. Lift your relaxed lower leg off the ground with your thigh muscles. Allow the knee and ankle of one leg to bend forward with gravity and you can place (or not place, if you prefer) your other foot down with complete control. Let the muscles, not the joints, absorb the weight as the foot touches the ground.

Now practice these suggestions for five or ten years and you might get somewhere. Seriously, I have been practicing my walking for years and I find that I learn things constantly and that my "incurable" and "chronic" injuries are healing. Instead of getting less mobile with age, I am becoming more light on my feet and my overall balance and body integration seem to be improving. Amazing!

The most important thing about walking is to not be afraid of falling down. It is my observation that when people are afraid of falling down, they stiffen up and try to keep themselves from falling.

Well, everybody falls down occasionally. If you fall down stiffly, you will generally hurt yourself more than if you fell down in a relaxed manner. This fear of falling becomes an unconscious part of standing and walking. Later in this section we will discuss *ukemi*. Ukemi is the ability to avoid being injured when hit or thrown, or even when falling down by mistake. This is a fundamental, and absolutely necessary, skill for the warrior. I was told once that a person will never be able to stand or walk in a natural and relaxed manner until he loses the fear of falling. Until then, he will be continuously and unconsciously forcing himself to stay up.

JUNAN TAISO

In this section we will discuss several specific exercises that can be used as a method of maintaining, or returning to the natural, full flexibility of your body. Junan taiso is also used as a warm up for taijutsu training. Simply stated, taiso is a series of stretching exercises with breathing, something like yoga. One of the added benefits of the

KEEP THIS CARD IN YOUR MOTOR VEHICLE WHILE IN OPERATION

IF YOU HAVE AN ACCIDENT- NOTIFY POLICE IMMEDIATELY

1. Write down names, addresses, telephone numbers and license numbers of persons involved and of witnesses.
2. Notify State Farm Agent promptly. (If any injuries, phone nearest State Farm Agent or Claim Office - If necessary, call information in nearest large town.)
3. Do not admit fault, do not discuss the accident with anyone except State Farm or Police.

HOW TO IDENTIFY YOUR COVERAGES

SEE POLICY FOR FULL NAME AND DEFINITION

A Liability
C Medical Payments
D Comprehensive
F Collision 80%
G Collision
L Physical Damage
H Emergency Road Service

R Loss of Use
R1 Loss of Use and Travel Expense
U Uninsured Motor Vehicle
S Death, Dismemberment and Loss of Sight
Z Loss of Earnings

KEEP THIS COPY IN YOUR CAR

STATE FARM INSURANCE

GEORGIA LIABILITY INSURANCE IDENTIFICATION CARD

POLICY NUMBER 164 3368-F15-11A

INSURED
PHILLIPS, LEWIS A

ORIGINAL ISSUE DATE DEC-15-68 EXPIRATION DATE JUN-15-93

CAR-YEAR/MAKE/VEHICLE IDENTIFICATION NUMBER
83 DODGE ARIES STA WAG
1B3BD59C1DF256112

COVERAGES (SEE REVERSE FOR COVERAGE NAMES)
A BODILY INJURY/PROPERTY DAMAGE LIABILITY
C MEDICAL PAYMENTS
D COMPREHENSIVE
H, U-250

AGENT STEVE STEPHENS
5105 PEACHTREE IND
BLVD SUITE 7 30341
PO BOX 80583
CHAMBLEE, GA 30366-0583
PHONE # 404-457-0209

training is that it will help you learn to breathe easily and naturally as you move. Breathing, as a bodily function, cannot be separated from movement. There is an intrinsically easy way of breathing that results from proper integration with movement. Even breathing at rest will seem more fun after junan taiso training.

In modern times, this form of flexibility and breathing training can be one of the most important things that we can do to promote good health and longevity. It increases oxygen intake and blood circulation, strengthens our bones, and helps us overcome the body's tendency to slump or twist, and then freeze into postures that cause us to constantly battle with gravity. It gives us energy by stimulating certain parts of our nervous system which in turn causes the production of a hormone called endorphin, which is a kind of natural healing drug and pain killer.

Junan taiso also affects one mentally. It allows tension to dissipate; we feel that we can slow down, take time to get to know our bodies better, or reflect on things we usually don't have time to think about. Its relaxing!

When performing junan taiso exercises, take it easy. Slow down, do not push past the verge of tolerable discomfort. Breathe, breathe, breathe. The term "no pain, no gain" has been taken to the extreme. Your body needs to feel a certain amount of stretch so that it learns its limits. This discomfort also signals for the release of that endorphin hormone. Remember, however, if you injure yourself, you will not be able to appreciate your natural range of flexibility. If the pain becomes too great, if you cannot smile as you are training, if it stops feeling good, *ease off!*

Flexibility is improved by consistency, not trauma. You must stretch every day, but only enough to experiment with your limits. Not only will your body feel better, but the physical act of stretching is a reminder that we should flex our hearts and minds a little each day also. Stretching is a good habit to get into; flexibility, a good frame of mind to be in.

The exercises that follow are not strictly "ninja-type"

Feel your connection to the earth

exercises; similar ones are used in other disciplines as well. These particular exercises are specifically recommended by Dr. Hatsumi and other senior teachers of the art. Although various junan taiso exercises have been presented responsibly in other books, the set pictured and described below are a recommended daily routine of nine exercises that cover all basic muscle and joint groups. They take less than thirty minutes to perform and are drawn from a much larger group of worthwhile exercises. Appropriate suggestions on breathing are included in the descriptions.

To begin, stand easy with your feet shoulder-width apart, hands hanging easily at your side. Take a few seconds and feel your connection to the earth.

Exercise One

Take a deep breathe, hold it, and let the bones and muscles relax around that ball of breath. Take a few more seconds to see if you feel totally and naturally upright. Quietly let the breath go. Do this three times.

Exercise Two

Reach up and out with your arms as far as you can as you breathe in. Breathe out, deeply, from the base of your stomach, pushing your hands together and forward. Be sure that your knees are relaxed. Feel as if your whole body is required to perform this movement. Do this three times.

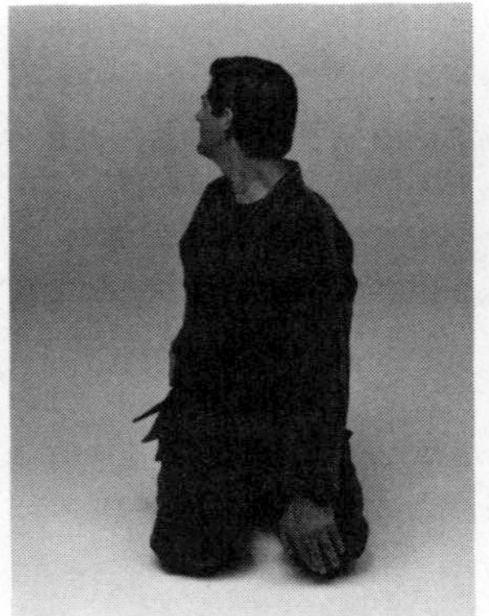

Exercise Three

Kneel down in what the Japanese call *seiza*. The back is naturally straight (repeat exercise one while kneeling if necessary). Your weight should be equally supported along the line extending from the knees, down the shin, and down the ankle to the toes. Take a deep breath. As you let the breath go, twist from the waist, through the shoulders and neck, feeling the stretch through the entire upper torso. When all the breath is gone, breathe back in slowly, returning to the starting position. Do this in all four directions as shown. Repeat the entire process three times.

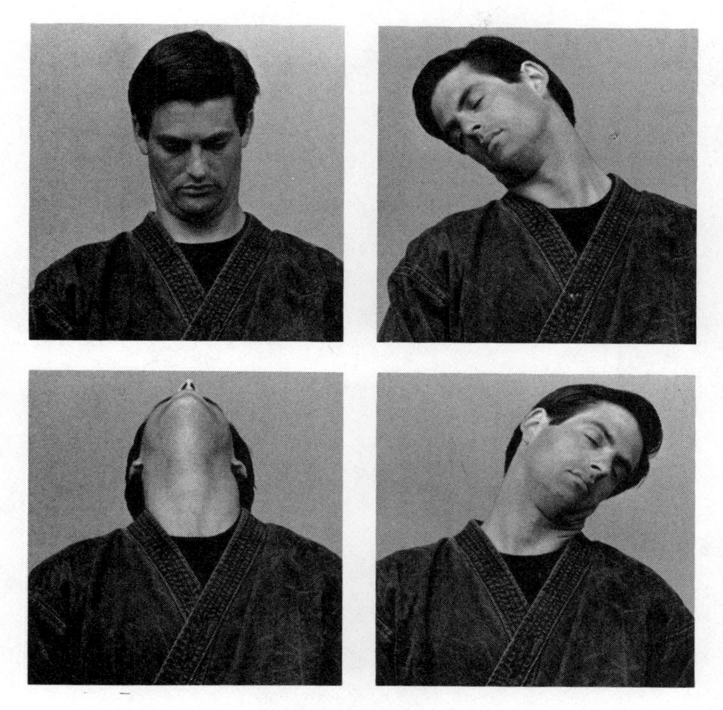

Exercise Four

- Rotate your neck to the right nine times, then to the left nine times. Remember to breathe and don't scrunch your shoulders. Your head should feel "heavy."

- Rotate your shoulders as you relax your arms. Do this nine times (or more) also. Feel the blood as it seems to pool in your fingers.

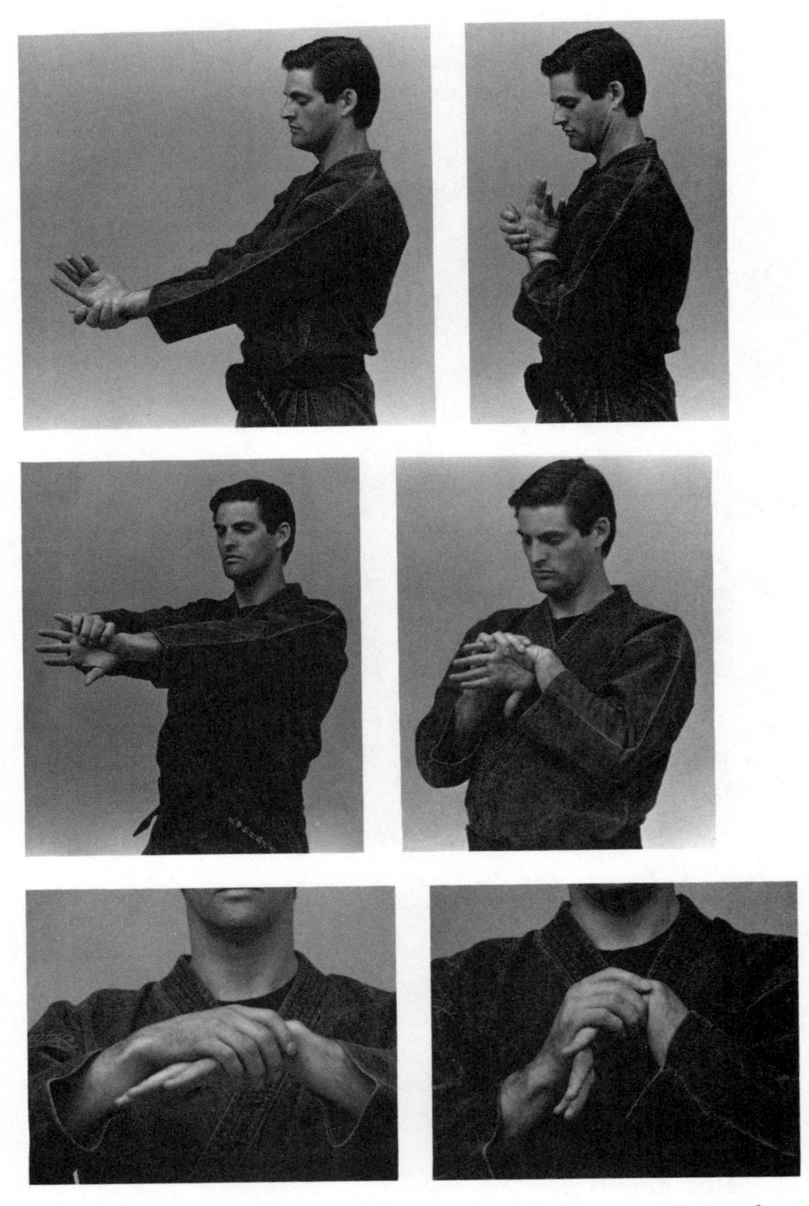

- Twist your wrists to the inside, to the outside, and then straight down, three times with each wrist. Relax and breathe out as you apply pressure.

- Stand up, spread your legs apart, and place your hands on your knees. Take a deep breath. As you let the breath go, twist your entire upper body to the limit. You should feel a stretch in the hips, back, and neck. When the breath is gone, relax back to the starting position as you breathe back in. Do this an equal number of times to the right and left.

- Put your legs together and roll your knees nine times to the left and nine times to the right. Keep your weight over your legs.

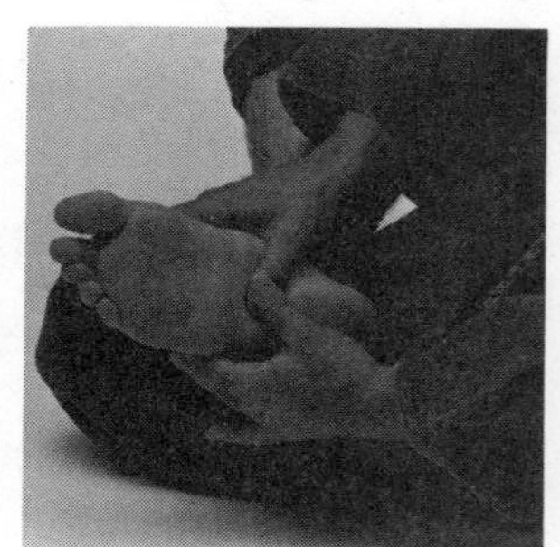

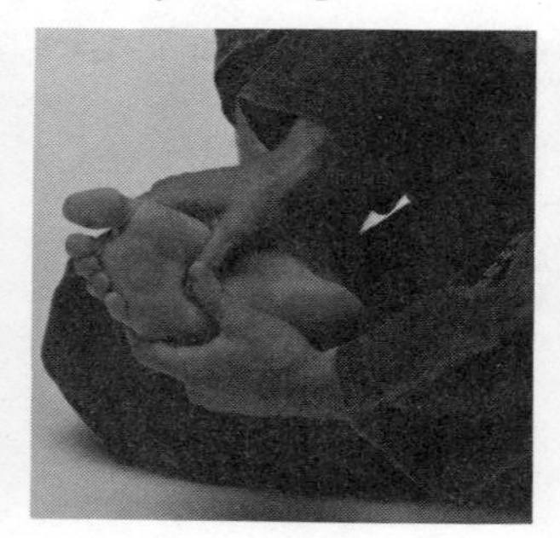

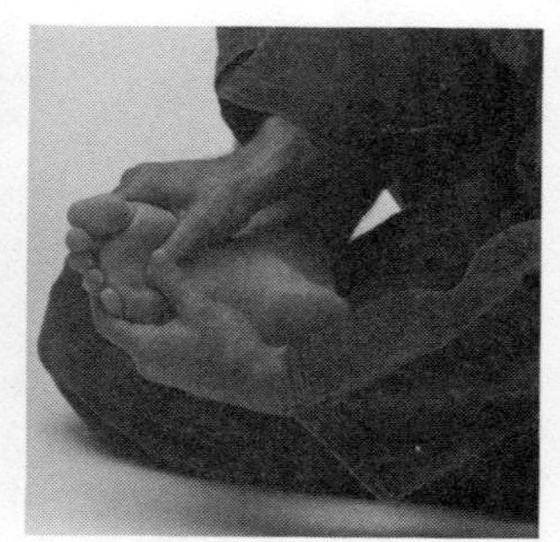

Photos continue on next page

- Sit back down, we are about to do the most important exercise of all. Cross your legs and take a foot in both hands. Massage the bottom thoroughly, toward the toes. Rotate, first the big toe, then all the toes, and then the entire foot at the ankle, nine times (or more) to the right and then the left. Do

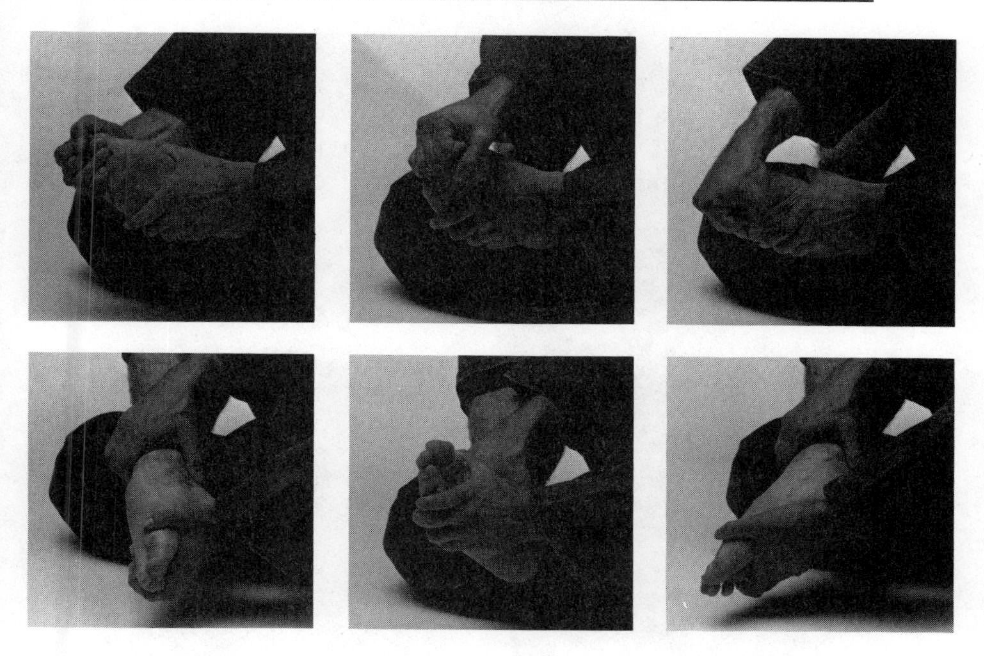

the same with the other foot. A Ninja's feet are very important. Take care of them. Don't let them get cold if possible, and massage them every day.*

Right about here is where your ego or impatience will get you in trouble. The next four exercises are traditionally oriental. I have found that, since most Westerners don't sit in seiza from childhood, they don't have the knee or lower back strength or the flexibility to perform these exercises as well as a typical Japanese. If you feel that you are having difficulty with any of them, refer to the optional exercise that follows each of the four. Use the optional ones to work up to, or, at least loosen up for, the performance of these

*A nice thing about this "rotation" exercise group is that you can do these exercises all through the day as you sit in the office, watch a football game, or wait in the car at a red light. Use your imagination and common sense.

exercises in the traditional way. I do. There is no sense injuring yourself. Stretch ONLY until you feel a slight strain. You have to be relaxed and you have to breathe, or you are WASTING YOUR TIME. You will not become more flexible and you will probably injure yourself. I have done that too.

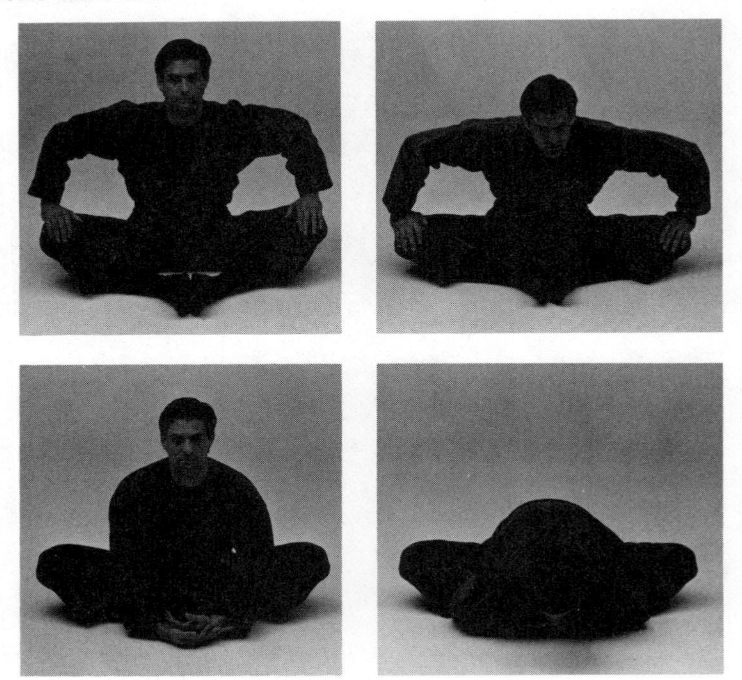

Exercise Five

Sit down on the floor, put the soles of your feet together and pull them toward you. Put your hands on your knees and *gently* bounce them up and down a few times to loosen up the joints and the muscles. Now, grab your ankles, feet or big toes and, keeping your back straight, press your chest toward your feet and your knees toward the ground, breathing out as you go. Be sure to hold your head in the same position as you do when standing normally, and don't put too much stress on the ankles. Relax off of the stretch a bit, as you need to take another breath, then repeat for a total of nine repetitions.

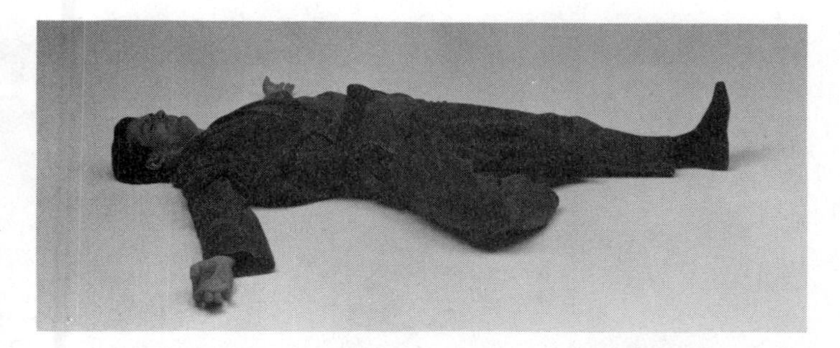

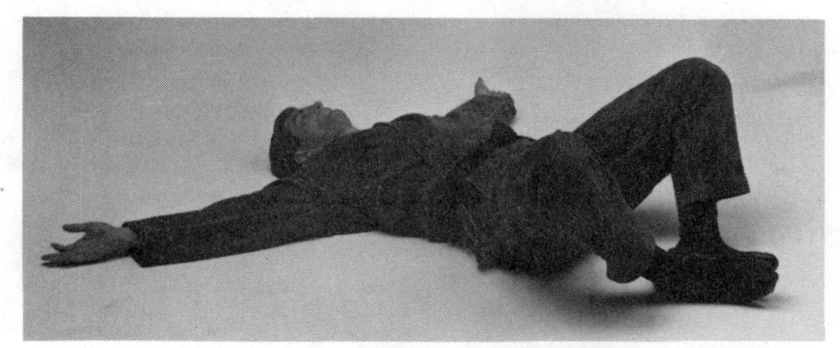

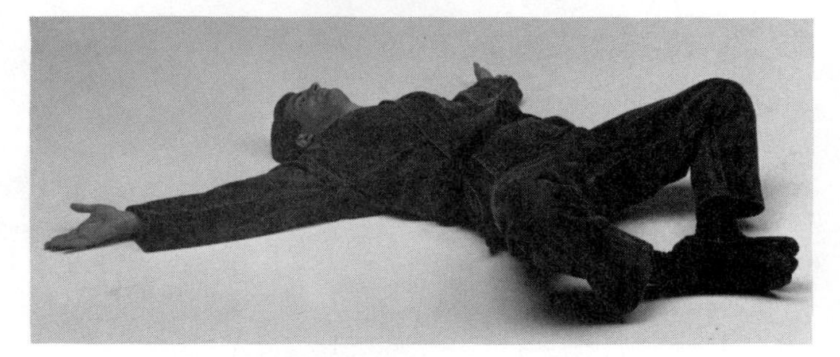

Optional Exercise Five

Lie on the ground with the small of your back flat against the ground. Pull one foot at a time toward the crotch, feeling the stretch in the hips. Eventually, pull both feet in together and *gently* bounce your knees toward the ground.

This optional exercise allows the legs and knees to stretch without your whole weight on them. It also protects you from the undue stress that will be translated to your lower back should you attempt the traditional exercise without loosening up the hips and knees first. You will see that all of the optional exercises are designed with this kind of safety in mind. Nothing makes you feel more stupid than hurting yourself while learning how to protect yourself.

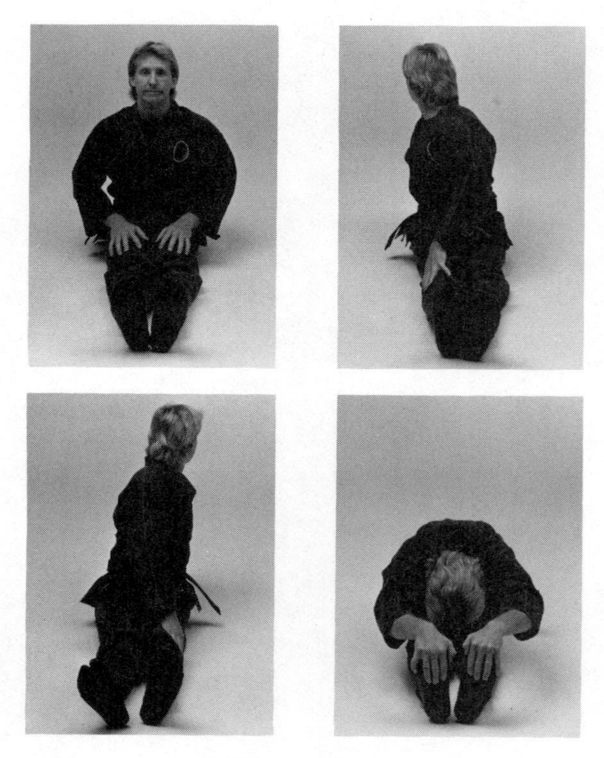

Exercise Six

Stretch your feet out in front of you, toes curled slightly toward you. Twist to the right and twist to the left a few times as you breathe out and relax. Take a deep breath, reach out and grab your feet or big toes and pull them toward you as you slowly breathe out. Press forward and try to put your chest on your legs, remembering to keep your head held up normally. When the breath is gone, relax off the stretch and breathe back in. Do this for a total of nine repetitions.

Optional Exercise Six

Lie flat on your back with your knees slightly bent. Lift one leg up and grasp it with both hands, feeling a good stretch. Gently lift your torso slightly off of the ground, in as natural a way as possible, to increase the flex. Repeat with the other leg. Breathe with all movements and don't strain. If you can't remain relaxed you are trying too hard. Go for quality of movement and a good feeling, rather than excessive stretch.

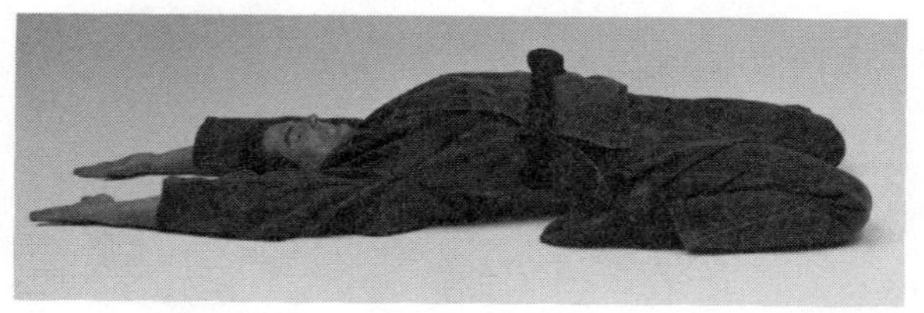

Exercise Seven

Go back to the seiza position and kneel straight up until your legs form a 90-degree angle. Reach back and put a thumb into the sole of each foot. Push your hips up and out. You should feel a slight stretch in your thighs and relief in the lower back. Stay that way for nine good, slow breaths, if you can. Now lower yourself back slowly until you are laying on the floor.

If you cannot go all the way, go as far as you can without overstraining. You may have to open your knees a little bit so that you can lie down between them. If this causes any strain on your knees, DON'T DO IT. Lie right back over your feet instead. Stretch your arms as far back, then as far to the sides as they will go. Move them all around if you like. Take a deep breath. As you breathe out, gently push your buttocks up off the floor. As you breathe out, gently lower them. Repeat for a total of nine repetitions.

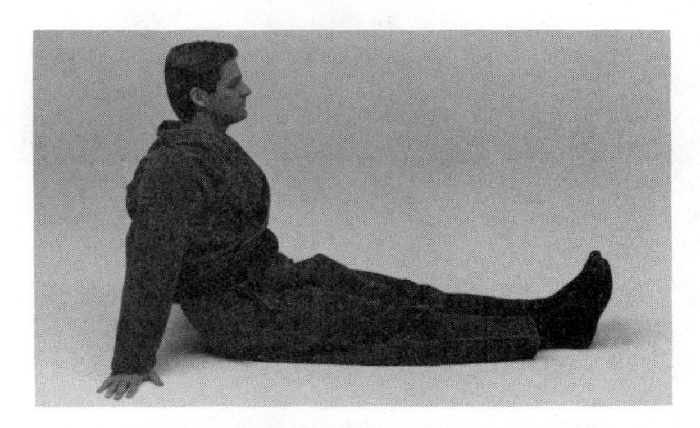

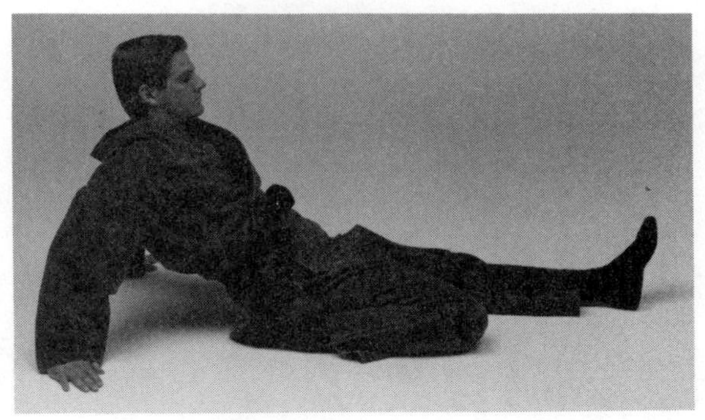

Optional Exercise Seven

Sit with your feet straight out in front of you. Bend one knee and gently try to lean all the way back. Go only as far as you can go without excess discomfort. Sit back up and repeat to the other leg.

Exercise Eight

Sit down with your feet straight out in front of you, then spread them wide, as far as you can. Massage, and gently pound the insides of your legs briefly to loosen them up. Reach out for your feet or big toes, pull them back toward you, breathe out, and lower your chest to the ground. Again, keep your back straight and your head held normally. Relax off the stretch as you need to breathe back in. Repeat this exercise for nine long breaths.

Optional Exercise Eight

Lie flat on your back. Reach between your legs and grab your big toes with each hand. Straighten one leg out and to the side, relax back in and do the other one. Eventually you should be able to do both at the same time, while holding the heels rather than the big toe. Work up to it gradually.

Very good. Now let's do the last one. Take your time, because it is a warmdown exercise as well as a stretch. Here is a hint on how to get the most out of it: It is easy to see how to utilize gravity to go down; but, how do you use gravity to get back up?

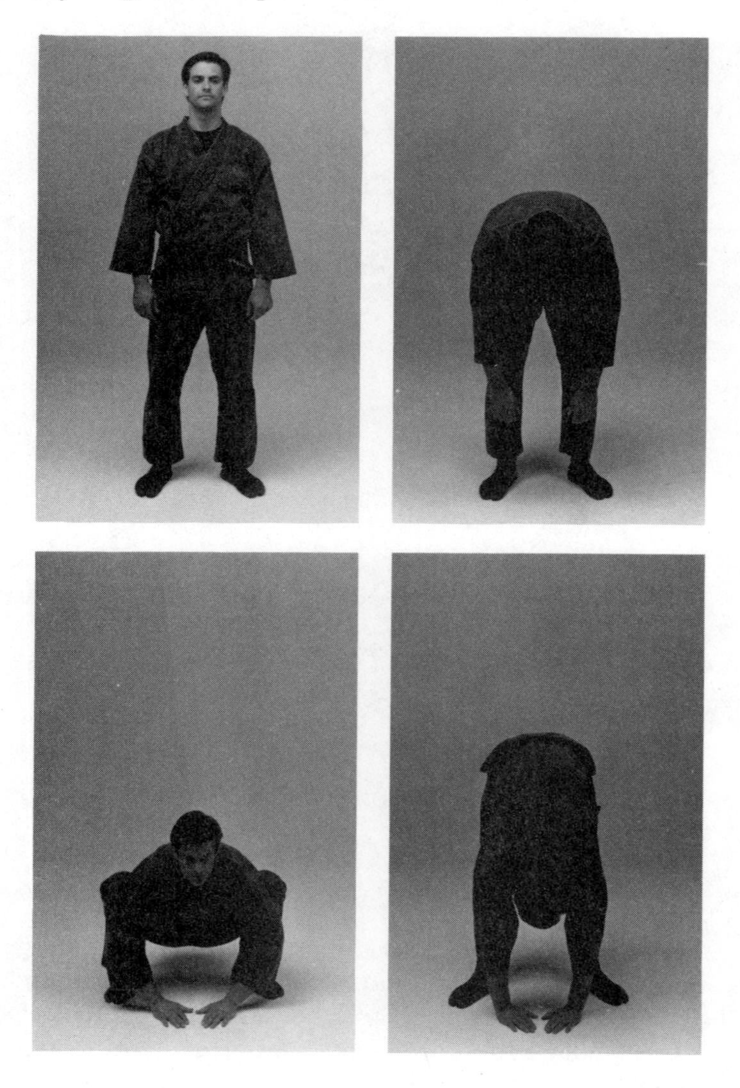

Exercise Nine

Stand up with your feet shoulder width apart and knees straight but not locked. Allowing your back to remain as natural as possible, slowly lower your hands to the floor. Go down slowly, breathe in, breathe out, and lower, breathe in, breathe out; and go lower. When you are hanging at your limit, relax your knees and go all the way to a squatting position with both feet flat on the floor. Return to the standing position using the reverse process. Keep your palms on the floor if possible. Take your time and breathe, breathe, breathe.

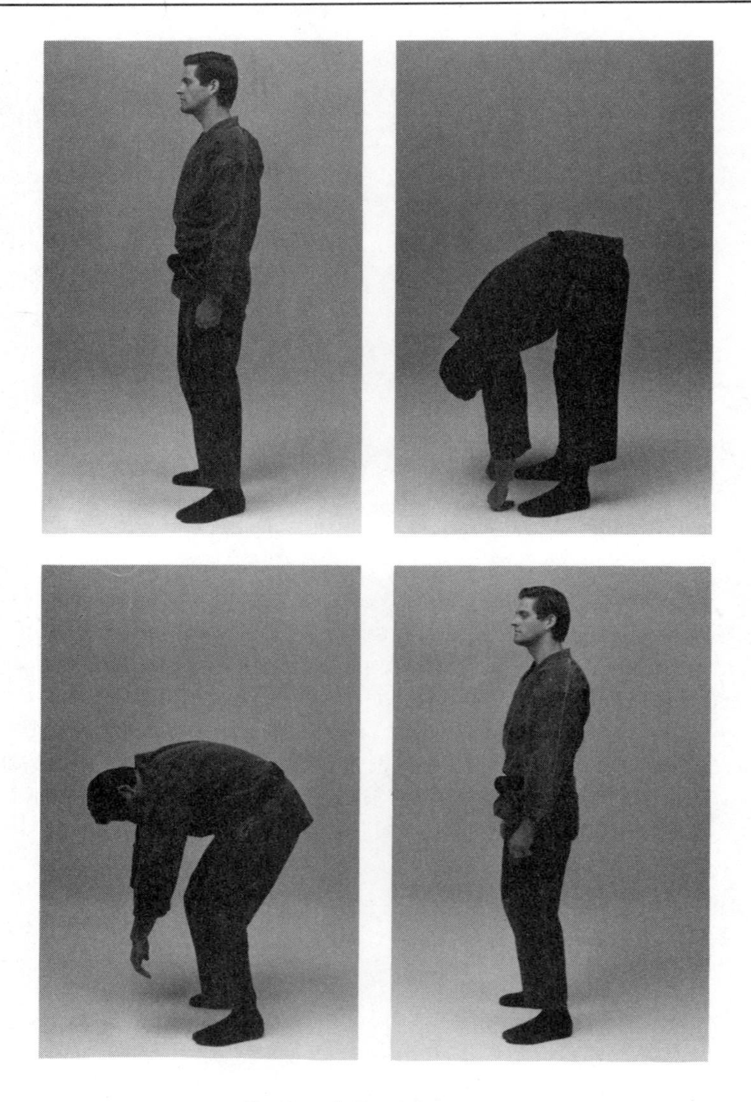

Optional Exercise Nine

Go for the floor, as before, and remember not to strain. When you reach your limit, hang there for a minute or two and feel your body weight. Bend your knees a bit and roll back up as you breathe. Can you find the way to move without straining your back?

As mentioned before, there are many other good exercises and stretches that you may discover that will help you gain the freedom of movement that you require. I do not think of junan taiso as muscle stretching, I think of it as a process that allows me to return to the complete freedom of movement that I had as a baby. Often it is the joints that lack a full range of movement, not the muscles that need "stretching."

Use junan taiso to discover your full range and freedom of movement. Remember to breathe, relax, take it slow, and keep your body as naturally as possible as you move through the exercises. A good routine is to walk for an hour each day and stretch for twenty or thirty minutes before you go to bed.

TAIJUTSU

Ninpo taijutsu is a method of moving the body. It is not the only method. Everybody who uses his or her body uses taijutsu: Hula hoopers, belly dancers, pizza tossers—all have a unique style of moving their bodies. Most of us have a method which is more nondescript.

Ninja have ninpo taijutsu. This is a very broad concept. It encompasses everything from the physical mechanics of moving, to a philosophy of moving through life. Physically, a ninja moves stealthily, quietly, without unnecessary effort, and *always* naturally. Philosophically, he acts stealthily on behalf of his community, family, and self. He moves toward a quiet justice that plays itself out naturally in the scheme of the larger picture. He never fights when he doesn't have to, he hides when the forces against him are too great. He will win without revealing his victory. He is a patient hero who does what needs to be done without fanfare and testimonials.

He can live anywhere and play any part. He will never lose his sense of justice or his presituational values in any case. He is a man of action.

These descriptions, of course, pertain to my romantic

idea of the ninja. You may see the value of true and heroic action in a different sense. Yet, any description of a hero contains the element of action. The following section describes the method that I use to develop my ability to move, and thus, to act.

UKEMI

The ability to avoid being injured when you are falling or being hit is called *ukemi.* One might immediately see the possible connection between the ability to stay safe when physically attacked, and learning to absorb emotional or psychic attacks as well. Let me state emphatically, however, that ukemi is a physical concept. The ninja does not learn ukemi in order to become inspired to persevere when attacked by words or thoughts. Rather, he is not harmed by emotional or even spiritual attacks *because* he understands ukemi on the physical level.

I will say that ukemi may be the most important of the taijutsu skills. If you cannot be hurt, there is no need for any other defense. For the ninja, the ability to persevere is paramount. Until one enjoys the process of being attacked as a lesson to be learned, he will never understand ukemi. Until one throws away the difference between "winning" and "losing" and replaces it with "keep going," he will always become injured when attacked.

Even the greatest warrior will die, maybe due to his own overblown sense of infallibility, maybe due to the luck of someone less skilled, maybe as a pawn in a larger movement of man or nature, or finally, of old age. There is no technique that will work in every instance. The most important thing is to die happy. There is a fundamental difference between a yogi and a warrior. Both understand life enough to lose their fear of death. Both can leave this world with a smile on their lips. But, the warrior has the power to defend himself and prolong the laughter a little longer. Sometimes.

This is one of the meanings of ukemi: If you do not care about dying, then you will not resist the bullet that is shot at you. Therefore you may be pushed out of the way by the force of the bullet, or the bullet might pass through you without doing enough damage to kill you.

Then again, the bullet may hit a vital organ and kill you anyway.

The way that I know to learn this ability is to practice moving my body with the force of an attack and by learning to fall without getting hurt. I was once told by a friend, Doron Navon, that I would never know how to stand up until I learned to fall down. Think about it.

Ground Hitting

There are an infinite number of ways that you may end up hitting the ground. Until you have trained yourself to the point that you consider the ground an ally and actually enjoy hitting the ground, you will almost invariably be injured. Please review the following set of pictures and remember to practice on a soft surface so that you do not become injured. Remember, ukemi training should teach how *not* to get injured.

Photos and text continue on next page

Start

Front roll. Get close to the ground and roll down one arm to the rounded shoulder, diagonally across the back to the opposite hip, down the leg to the side of the foot, and roll your foot from the outside to the ball and move gently to your feet.

Now watch a similar roll and escape, as the opponent tries to force the defender to the ground with a wrist twist.

Front roll used to escape from twisting wrist grab.

Back roll. This is the reverse of the front roll. Roll onto the side of the foot, up the leg to the hip, across the back to the opposite rounded shoulder, up the arm, and onto the feet.

On the next page is a similar roll in the context of a different kind of wrist attack.

Back roll used to escape from twisting wrist grab.

Side roll. This is for moving—you guessed it—sideways. Slide one leg across in front of you while lowering yourself straight down on your other leg until your buttock is on the ground, roll up your same side and from one shoulder to the next, and then up onto your feet.

Here is a variation of the side roll to escape a punch to the face. Notice how the leg is used to attack the shin of the assailant.

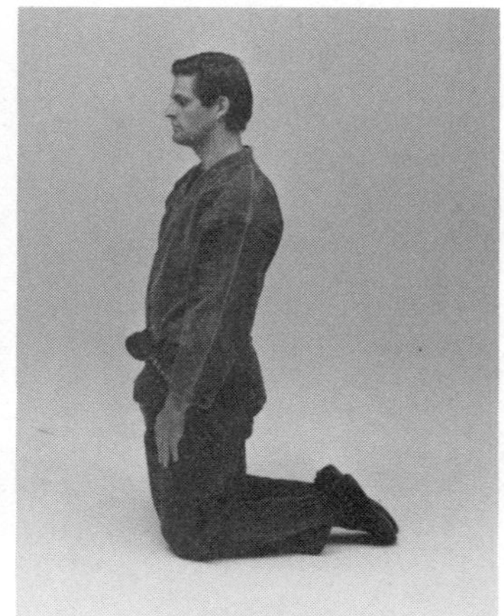

Body drop. Kneeling upright, allow yourself to fall on the meaty part of your lower arm (attempt this on an extremely soft surface until you are sure of where that meaty part is). Arms are at about 60 degrees, hands are face high.

Eventually you can try this from the standing position, practice using one leg or the other as a counterweight, as you fall. Keep your body naturally straight. Try not to bend at the waist.

There are many other ways of falling without hurting yourself. Practice these until you enjoy them (this may literally take five or ten years). Others will come to you with continued training, and most importantly, in the heat of a truly dangerous fall.

I remember awkwardly slipping off of a wet platform and falling eight feet to a marble floor one time, only to roll lightly to my feet completely uninjured. I was the most surprised of all of the people who witnessed it. They couldn't believe I wasn't injured and looked at me very strangely. I wondered where I learned that particular breakfall, and realized that I had made it up on the spot. Lucky for me.

After reflection, though, I suppose that it is necessary to be able to take the building blocks of repetitious training in the fundamentals and use them to create a new and unique solution to a new and unique problem.

The only way to get better at rolling, and to give yourself the ability to spontaneously create appropriate techniques, is to roll. Try to enjoy it and you will improve quickly. Many people consider break falls to be the distasteful part of their training and hurry through it with a grimace on their faces. They will never understand ukemi, and therefore, they will never understand Ninpo.

Here is a hint: Control your fall by bending your ankles, knees and hips. If you review the pictures, you will notice that the defender is quite close to the ground before he actually falls. If your joints are inflexible, on the other hand, it forces you to fall from much higher.

Absorbing Strikes

The ability to absorb a strike is critical. I don't care who you are, the possibility of getting hit in a fracas of some type is very real. It might just be a lucky punch thrown by someone less skilled than you are, or you might be struck by a car. In either case, having a feel for ukemi might help you save yourself from serious injury. In the following two

instances a punch is thrown in one, and a harsh shove is committed in the other.

Here, in the first, the punch is absorbed by movement of the legs. The face actually catches the punch and the body is responsive enough to move the face back along the line of attack until the distancing is such that the punch is no longer dangerous. This movement is sometimes referred to as "rolling with the punch." I think the word "absorption" describes it more accurately.

It takes many years to develop this skill to perfection. A good way to practice it, of course, is in slow motion. Have your partner slowly punch you while you try, gradually, to pick up his rhythm, so that when the punch hits your face you and his fist are moving at about the same speed. Too fast is no good, he'll see you move and change his attack.

This just starts your problems all over again. Too slow is no good either, for obvious reasons.

Again, start slow and work your way up, gradually, to a realistic speed. Here is an image that may help: Picture yourself watching a rotating merry-go-round. How would you get on it now that it is already moving? You can't just jump on, you'll get thrown around. Try walking with it, faster and faster, until you and it are moving at the same speed. Now just calmly side-step up. You're on!

In the second example, one person shoves the other. The defender absorbs the shove by spinning on an imaginary axis that runs through his body. There is little or no resistance, so there is little or no pain.

The secret to this movement is not to resist. As you practice this exercise you may get the tendency to do one of two things: Fight against the shove, or try to anticipate the shove and move before it hits you. Refrain from either tendency—don't *care* so much and you will find that you can absorb the shove quite easily.

THE FIVE STRATEGIES

Now that you have been introduced to methods of avoiding injury, I would like to begin talking about the sphere of action. In ninpo taijutsu we sometimes refer to five basic attitudes. The parallels between these attitudes as manifested in physical confrontations, and how they might apply to emotional and spiritual confrontations, will be

EARTH

(chi no waza)

WATER

(sui no waza)

obvious. Again, however, they must be understood from their physical nature in order for them to be useful. Without the physical foundation, all of these theories become reduced to intellectual curiosities.

Collectively, they are referred to as the *godai,* and they represent five possible strategies in a fight. They are used as distillates of the myriad possible feelings that one might actually have in a confrontation, and they are used as a training aid only.

The problem for non-Japanese who are trying to utilize the Bujinkan training method is that the attitudes are coded according to sets of oriental philosophical concepts. For example, the godai consists of the set: Earth, wind, fire, water, and void. If you have read any of my other books, you know that I associate the earth element with a standing firm strategy, the wind with an avoidance strategy, the fire with a committed strategy, and the water with

FIRE

(ka no waza)

a responsive strategy. The void, I characterize as the strategy of no strategy, or the spontaneous action of a trained warrior as he responds to true danger.

For a Japanese, the use of this set, connotatively, feels comfortable from childhood. Westerners have to learn the connotations. Due to the cultural gap, I am not sure that we ever quite do. I invite you to study the four Japanese exercises below, which pertain to each of the first four attitudes. We'll save the fifth for more discussion before illustrating it.

WIND

(fu no waza)

I congratulate you if you can sense the elemental difference in each, and how each uniquely represents a particular strategy. After some years of thinking about this, I have come to the conclusion that the feeling of the strategy is in the person, *not* in these particular exercises. The exercises are merely one way to allow that inner feeling to manifest itself physically.

The problem for us is: What if we don't already have the feeling culturally implanted in us from childhood? We might merely imitate the physical shell of the exercise and never truly understand its essence.

It is very important to be able to assume each of the inner attitudes as you train. Therefore, you must find a way to get the feeling of each, if not the entire cultural context. Fortunately for us, Stephen Hayes created a series of exercises that does just that. I think it is one of the most brilliant devices that Western man has ever created to help himself understand the oriental thought process. The reason that the exercises are so great is because they allow the practitioner to feel, in a physical sense, the essence of the attitude. Many Western writers have attempted this before, but their explanations are all intellectual, and therefore, remote. Try the following five exercises and internalize the feeling that they help you evoke.

Earth

In the earth feeling, you are aware of your right to live your life unmolested. The strategy manifests itself in a standing firm attitude. Your body is solid, naturally erect, and your feet are shoulder width apart. You feel deeply rooted in the ground and also in your conviction that you can handle any situation that comes up. Your breath is deep and seems to come from the buttocks. Maintain this feeling for a minute or so, and then ask a training partner to slowly swing at your head. Calmly lift your hand in a "stop" gesture and let him try to move you. If you have the whole body attitude (conviction, breath, and physical attitude together), you will find that it is very easy to withstand the hardest strike with ease.

Now go back and practice the Japanese chi no waza. Remember to maintain the standing firm attitude, even though you are moving through the exercise. Put the feeling from the Hayes exercise into the Japanese movement.

Water

In the water feeling, you have arrived at the realization that you must respond to the attack. All humans are incomplete, and the fact that you may not be strong enough, fast enough, or well-enough prepared for the attack is normal and natural. Imagine the queasy feeling in the pit of your stomach that fear tends to generate, and move back at a 45-degree angle, leading from that spot. Breathe from that spot, also. As you do this, you should have the same feeling as you would have if you were to attempt to sit down on a chair that is suddenly pulled away. Your back foot doesn't "step" to where it ends up. It is brought back so that you don't fall.

Have a training partner try to grab you, and use this same smooth, back-pedalling movement to glide back and away from his grasp.

Again, keep the responsive feeling, and try the sui no waza. It is the feeling, not the movement itself, that defines the strategy.

Fire

In the fire attitude, you have a joyous certainty as to what must be done. Philosophically, the joy is based on the "doing of what is right." I have presituationally defined what that might mean in the first two sections of this book. The movement looks like an exaggerated handshake. Move and breathe from the solar plexus, and see if you can assume a totally committed attitude.

Now, as your training partner goes to hit you, you easily sweep up and shove him back, overcoming his attack with a committed defense.

This is one of the Japanese exercises that lends itself most easily to Western interpretation. You should have no problem maintaining the committed attitude as you practice the ka no waza.

Wind

The wind attitude is very ninja-like in that it is very patient and benevolent. You are avoiding the confrontation, either entirely or until the time when a counterattack can be performed easily and safely. To begin, picture yourself as extremely non-confrontational and light on your feet. Even your breath is shallow, and centered high up in your body, near the throat. Imagine that you can move your whole body from your ankles, and that the rest of the body will naturally respond to the movement.

As your attacker punches to your head, shift to one side or the other, maintaining this avoidance attitude. Go for the feeling. Don't spend a lot of time wondering whether this exercise is tactically complete. It is not meant to be. None of the preceding exercises are fighting techniques, per se. In proper context, they could be *part* of a defense strategy, but that is not the point here.

Now, maintain the feeling of lightness and avoidance as you practice the fu no waza.

Void

This is a tough one. The void is an absence of any particular, or easily definable strategy. It is my opinion that it is nearly impossible to operate safely in the void when one is in a learning mode. In other words, it is improper to just "go for it" and unthinkingly "see what happens" while you are training. People get injured that way, and besides, the untrained person will probably revert to the use of the low-level methods that he would have used in his pre-training days. If you have ever seen beginner or intermediate martial arts students sparring, you will know what I mean. They use none of their training, and the exercise usually turns into a brawl.

Occasionally, while training, I have felt myself move from the void. It is a great feeling, but I can't make myself move from there all of the time. I don't even try. I usually train in context of the training method: Using the fundamental techniques and variations to improve my skills. Hatsumi sensei feels that if I approach the training in this manner, I will move in the void realm more and more. I watch him move and I sense that he is always "just going for it." Even though he doesn't appear to be thinking about what he is doing, however, the results always work; and his movements always seem to conform to the ninpo taijutsu philosophy.

VOID

His movements remind me of a great musician. If you have ever listened to a great jazz artist, he seems to be making up a melody as he goes along. He is nearly always playing in the right "key." But the notes always come out as a unique expression in the context of the music. Obviously the musician is not thinking about each note that he is playing. It is amazing to think that this creative ability has at its basis many years of playing boring scales.

I think of the godai, and the kihon happo exercises that will follow shortly, as the notes or scales of taijutsu. With years of practice I hope to be able to use them spontaneously and creatively, like an artist, to come up with solutions to unique self-defense situations. I suppose, however, for now I will just have to "keep going."

In the meantime, I believe that the earth, water, fire, and wind attitudes are ones that we can use to clarify our strategies in training. It is the void attitude, however, that we will be in, automatically, in a real confrontation. If you have ever been in a fight, or more likely a car crash, you will recall that your emotions in that hyper-concentrated, seemingly slow-motion period are extremely hard to categorize.

Obviously, there is no specific attitude for the void. As a result, both the Japanese and Stephen Hayes were at the liberty to introduce any concept they wished as a key to

understanding the void and its uniqueness in terms of the other, more easily definable, attitudes.

Mr. Hayes chose the opportunity to present ukemi, which has already been covered in this section. The Japanese chose to present the concept of *kyojitsu tenkan ho,* or the juxtaposition of truth and falsehood. Kyojitsu has deep philosophical and psychological implications that cannot be adequately covered in this book. Basically we are talking about the ability to present a countenance that appears to mean one thing, but which really masks another.

I realize that this is not a very satisfying explanation. Unfortunately, kyojitsu is just that elusive of a subject. It is introduced in the Japanese version of the Godai as a feint or distraction. The specific method illustrated is a form of *metsubishi,* or blinding technique. As you will see, it creates the illusion that the attack will come from the upper body by drawing attention to the hand and away from the kicking leg, which actually performs the counterattack.

This technique is a simple introduction to a very profound subject that may be at the very core of ninjutsu strategy. Just be aware that things may not always be as they seem. Have patience and subscribe to this ninja motto: For the ninja, there is no such thing as surprise. The application of the *ura,* or other side, of this admonition is the source of kyojitsu tenkan ho.

KIHON HAPPO

The kihon happo, or "all-direction fundamentals," are the inspirational source for the fighting method eventually employed by a ninja in combat. Actually the term, itself, has been rather misleading to the Westerner. In Japanese, the term *kihon* means fundamental, while the word *happo* is derived from the word *hachi,* which means "eight." When the "*ha*" becomes "*happo*" it ceases to mean eight and becomes "all-directional."Although the kihon happo is sometimes taught as a series of eight exercises (which further compounds the misunderstanding), it may be more useful to think of them as that set of all individually distinguishable techniques, be they kicks, punches, throws, or immobilizations, that we can practice to gain taijutsu proficiency.

When I first heard the term kihon happo, I was looking for eight techniques. However, when I asked the various senior teachers to show them to me, they each had a different version. Believe me, there are more than eight. Including variations, there are an infinite number of them.

Below, I have selected eight basic techniques from the kihon happo. Hatsumi sensei told me that, from each of the eight, I should try eight variations. And from each of the eight times eight, I should try eight more variations, and so forth. I am sure that you can understand where this all leads. We are talking about a training method that provides for infinite variation and flexibility. Indeed, as I have experienced, just when you think you have seen it all, there is another way to approach the technique. Maybe you will be the one that spontaneously discovers it. Who is to say how many other *henka*, or variations, lurk in the void. I have come to believe that there may be one for every self-protection circumstance, if only you are open to receive it.

TRAINING PHILOSOPHY

Before the illustrations of the specific techniques, I would like to discuss the basic philosophy of the training method that I employ, and how it differs from other training methods that I have tried.

Actually, this taijutsu training is a path to enlightenment. We have a training motto that is: *Shikin haramitsu dai komyo*. It means, roughly translated, that in each action there is a potential to find the enlightenment we seek. Since the warrior is a man of action, this motto is an apt one.

By doing my own research, I have tried to discover how this can be true. I looked at other, so called, paths to enlightenment, such as meditation. I even read about the effect of such diverse methods as hallucinogenic drugs and tantra, to find a parallel. This is what I have discovered: In each one, an unusual mind state is achieved that allows the practitioner access to a higher level of consciousness. In meditation, it comes as a result of prolonged contemplation.

With the use of drugs it comes as a result of a chemical reaction. In some methods of tantra it occurs at the moment of sexual climax.

In warrior training, I believe that it comes as a result of exposure to danger. It is at that moment when the fist is flying at our head that our mind is capable of opening up to a higher consciousness.

All of the other methods require a proper setting for best results. I believe that the setting is very important for warrior training, as well. Although I do not take drugs, I have read some of Timothy Leary's writings which attribute our country's overall bad experience with hallucinogenic drugs to the bizarre circumstances in which the drugs have been taken, rather than to the drug itself. This makes sense to me. If the mind is opened and then filled with crap, that is sure to cause bad results.

Of course it is silly to expose ourselves to real danger in the training hall without a proper setting, as well. Our training method requires an attacker and a defender who are working together toward the goal of increasing each others' skill. In short, one part of the setting consists of an attacker who can simulate the dynamics of a real attack with the intensity that can be handled by the defender. The other part of the setting requires the defender to have been taught tried-and-true methods of defending himself according to taijutsu principles. By using this formula, can you see how it is possible to create a dangerous situation in which the defender can still learn safely?

Generally, in ninpo taijutsu training, the principles of distance, timing, and rhythm are taught rather than those of speed and power. In real combat we are not concerned with the same things as we would be if we were engaged in a competition of some sort. We can assume that the antagonist, whether it be man, machine, or nature will not be matched to our weight class or any other sport-balancing factor.

When the objective use of our skills is our own defense or that of our loved ones, rather than demonstrated prowess in a contest, the entire nature of the conflict is different. Gone is the differentiation between winning and losing in a competitive sense. We do what we must do to maintain our life in the best way possible.

Another factor is the nature of our own actions. If the premises in the first two sections of this book are to be considered valid, then there is a fundamental rightness to our actions as they serve to defend our inalienable rights. Therefore our actions should be natural. If they are natural they should flow as easily and effortlessly as time itself. They should have the inexorable power of nature. Therefore, strength and speed in the usual sense should be unnecessary and, probably, incorrect—particularly if emphasized over the natural rhythm of the fight.

Allow this concept to become an extension of the dream that every warrior has: To be invincible in his rightness. You have the image of the ultimate warrior who defends himself totally without effort—you are just that powerful.

Now let's come back down to earth. To become the effortless warrior, does it make sense to practice groaning and sweating to accomplish your goals? Believe me, there is a lot of sweating and groaning that goes into the perfection of warrior skills. But it is the sweating and groaning that is a reminder of the unnaturalness that we are working through. In some training halls, it has become the means and the end.

I urge you to train with the goal of doing every technique without effort. Perform the movements in such a manner that will allow you to execute the techniques without speed or power. When you find that a particular technique does not work, study your movements and look for the flaws. Try the technique even more slowly until you catch the secret. Only *after* you find the secret should you ask your training partner to raise the level of training intensity.

One of the things that I have learned by practicing this way is that it is the rhythm, not the speed, of the attack that I have to match. Once I steal the attacker's rhythm, even at a very slow speed, I find that I can then perform a counter technique at any speed. If I don't have the rhythm, I find the counter to be awkward and difficult, even at a snail's pace.

This is why I recommend that you practice slowly and search for correctness in your movements. When you find the natural rhythm, and you attune yourself to the universal justice, you will act in concert with both.

As I said, you will experience pain, you will groan, and you will sweat as you train. But these side effects should not be considered any more a part of the finished product than the chippings are considered a part of a finished statue. The chippings are what the statue was not. Pare away all of the unnecessary effort and you will be left with a martial *art.*

Your training partner is a very important part of the method. Look for a friend who will help you learn about the nature of enlightened combat as you train. In other words, he must have the sensitivity to be capable of subjecting you to real danger at the exact level of intensity that you can just barely handle it. This is an unusual type of person. First, he must be non-competitive in the usual sense. He must also be able to simulate the dynamics of a real fight at any speed. This takes real talent. You will require your partner to attack you in the same way as you can expect to be attacked by a real assailant, *but at the speed that allows you to pick up the rhythm of the attack and successfully counter it.*

When you finally attempt the kihon happo techniques with a friend, keep in mind that you are each helping the other to learn.

KAMAE

In order to practice the Kihon Happo, you must begin your training from somewhere. This starting point is referred to as *kamae.* Kamae is a very deep and difficult concept. I will do my best to give you a flavor of what it means to me. I will have to ask forgiveness in advance if, after reading the next few paragraphs, you are no more sure of the meaning of kamae than before I started. Perhaps it cannot be explained intellectually, and can only be understood when the practitioner has experienced the spiritual and physical elements of kamae, as well.

Kamae is a word that many people think of as a physical stance or posture. It is usually shown in pictures as a way a martial artist would stand while he is preparing to execute a technique. This may be misleading. I prefer to think of kamae as an attitude, or approach to moving with taijutsu. For those of us who have yet to totally internalize this art, it is a vital point of reference from which, and through which we move.

Kamae is a point of reference within a movement. Think of the analogy of trying to use a map: First you must know where you are. Unless you know that, how do you know how to get to where you want to go?

Kamae lets you move from a familiar place. It lets you say: "I am here." If, while in the middle of practicing a technique, I get confused about what I am doing, I assume an appropriate kamae. This "finding of oneself" in the technique seems to be a psychological, as well as physical reassurance. Not only do I feel more physically comfortable in kamae, the return to the familiar gives me great psychological encouragement as well. I get the feeling that, once again, I know where I am in the fight, and I know what to do.

Kamae is an awareness of self.

In some forms it may look to others that you are "just standing there," but it is far more than that. It is a physical manifestation of your presituational values. Think back to the discussions in the second section of this book. My kamae is the physical embodiment of my value system and my spirit. Kamae is the physical manifestation of all that is in you, as you wait for danger. It is the foundation of taijutsu, the bedrock on which you build your philosophy of movement.

Ichimonji no kamae

Jujimonji no kamae

Hicho no kamae (front and side view)

Kamae is an elusive subject. I do not remember when I went from "just standing there," to "observing from kamae." It happened gradually, over several years. All I know is, that at my present level, kamae is the basis for all of my movement. The eight kihon happo techniques that follow all contain kamae.

The first three are obvious: Number one is ichimonji no kamae. It is the basic or, as its name suggests, "first" attitude of Ninpo Taijutsu. Your feet are at a 90 degree angle, legs comfortably apart, weight mostly on the back leg. You should feel as if your center is in the lower abdomen. The front hand is a lure, floating gently out in front of you, pointed at the opponent's eyes. There should be no tension in your elbow, shoulder, or knees. You are waiting.

The ichimonji will appear constantly throughout the kihon happo, each time in a slightly different context.

The second is jumonji no kamae. It suggests a committed attitude, as in the ka no waza. The legs are shoulder width apart, one foot slightly ahead of the other. Arms are crossed. The self is centered in the chest area. You gaze steadily at your opponent.

The third is hicho no kamae. It is a very flexible attitude. A static photo series does not do it justice at all. Hips are open to a 90-degree angle. The front leg is drawn up under the knee of the back leg, which supports the weight. The front hand drapes across the body, in this example, hiding the fist of the other. The self is high in the body. The eyes are birdlike.

The kamae in the remaining five kihon presented in this volume are not as obvious, but they are certainly there. They are strong foundations for your movements and important sources of confidence for you as you train.

I hope that you enjoy practicing these exercises and that they become a vehicle for you to learn much about yourself. Remember that I have shown only one version or interpretation of each technique. There are an infinite number of other variations.

INTRODUCTION TO THE KIHON HAPPO

Each of the various techniques I will describe have a Japanese name that is sometimes associated with it. I have purposely not included them for two reasons. One, the techniques are sometimes called one thing and sometimes another. This is confusing to Japanese and non-Japanese speakers alike. Second, the names themselves are confusing. Sometimes they seem to be a literal description of the technique, and sometimes the name doesn't seem to have a thing at all to do with the technique! This may be because, over the years, certain techniques were given certain code names, or that the words or phrases used had a cultural connotation that is not literally apparent. I do not know the history of each of the technique names, so I will refrain from including them.

Kihon Happo Technique One

You are in ichimonji no kamae. As your partner throws a straight punch to your head, you move your right foot to the right, forming a 45-degree angle to the attacking arm, simultaneously striking the inside of that arm with your fist. When you move, be sure to move from the hips. Your movement must be integrated, all pieces of the body beginning and ending as one.

Now, move forward over the left knee, simultaneously adjusting your distance to the target by moving your feet, until you can step with the right and land a (palm up) *shuto* strike to the neck of your partner.

The previous technique has many variations. The one pictured is rather straightforward and affords an opportunity to learn many important lessons. One lesson is proper angling: Using a correct 45-degree angle is very important. Another is distance: It is *your* responsibility to make the distance right. When you hit with the shuto, you want your hand to be right on target at the moment of impact, yet

your body should be in the powerful kamae that is shown. Arms and torso should be naturally straight. As you move from the receiving to the countering kamae, move smoothly through the hips and keep your weight over your knees, and your knees over your feet. Don't bob up and down.

Kihon Happo Technique Two

You are in jumonji no kamae. Your training partner throws a punch at your head. You drop low in the hips and into a 45-degree angle to the attacking arm. As your right foot grounds, your left arm should strike the inside of the attacker's right arm simultaneously. Smoothly, rock forward on your hips over the front knee and deliver a *boshi ken* thumb strike to the ribs.

Finally, return to the jumonji no kamae as you obscure your partner's vision with your left hand.

This technique is rather difficult, in that you may have a tendency to lose your kamae. Concentrate on keeping your torso natural and supple. Don't bend at the waist. Work the distance with your feet until everything "fits." There is a subtle "reverse" power contained in the initial strike made as you seem to retreat. In this particular variation, the strike should land just as your body is grounding itself. In other words, the hand and the foot make contact at the same time. The body is naturally erect. With practice you should be able to handle a series of rights and lefts easily.

Kihon Happo Technique Three

You are in hicho no kamae. Your partner attempts an uppercut to your stomach. Lower your weight and strike the inside of his wrist with a *shikan ken* knuckle strike, and kick with the flat of your left foot into the ribs under the armpit. Move forward over the left knee and adjust your distance to the target by moving your feet until you can land a (palm down) shuto strike to your partner's neck.

This is a difficult variation that also contains many lessons. One is balance and flexibility. If you are stiff in the knees and ankles, this set of movements will be very difficult. Distancing is also a problem. After the kick, your partner may move closer, stay where he is, or move further away from you. Without thinking, you must be able to adjust the distance so that you will be able to hit with the shuto from a strong kamae. If, when you administer the shuto, you find that your opponent's neck is too far to reach without craning forward, or too close to hit without bending the arm or torso, *you* must remedy the distance problem by moving your whole body to where it is supposed to be.

Shizen no kamae

The remaining techniques start from the *shizen* no kamae, or "natural" attitude. The shizen no kamae *really* looks like you are "just standing there." You must, however, learn to maintain your awareness of self without a "stance." Start practicing now. There are plenty of more recognizable kamae, particularly the ichimonji no kamae, within the techniques. Look for them and use them to help you to remain comfortable within the movement.

Kihon Happo Technique Four

As you stand in shizen no kamae, your partner grabs your lapel with his left hand. Step back to your right at a 45 degree angle, covering his hand with your left hand as shown. This movement should take your partner slightly off balance. Rotate your body to the left which turns the opponent's wrist over and allows you to exert pressure on the elbow. Lean into the elbow and twist the wrist to defeat his grip on your lapel and force him to the ground.

This technique should take no power at all. If you feel that you are wrestling with the opponent, it is *you* who must move to a more advantageous position. Practice this, and all of the techniques slowly, so that you can search for the secret of performing the techniques effortlessly.

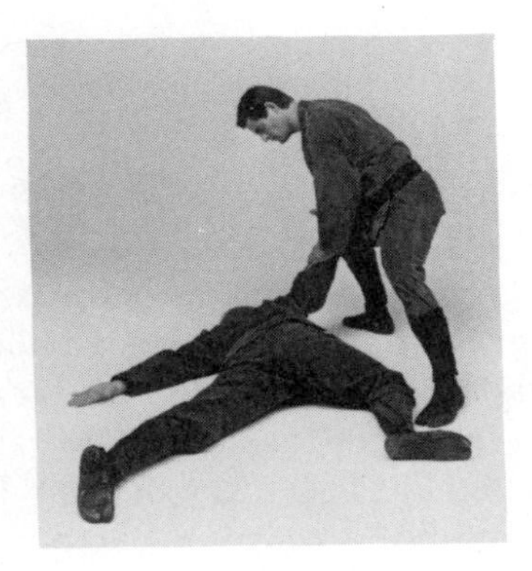

Kihon Happo Technique Five

Your partner grabs your lapel as before. Step back to the right at a 45-degree angle to break his balance. Rock back in over the right knee and snake your arm inside your partner's to trap the elbow as shown. Rotate your body to the left and kick his left leg to topple him. Put pressure on the shoulder to immobilize your partner once again by rotating your body.

There are two major secrets to this last technique. One is that you must break the attacker's balance by either angling or distancing. Second, use your body weight and taijutsu movement, not arm strength, to accomplish the immobilization. No "wrestling around" should be necessary. If there is, you are doing something wrong. Slow down and look for the secrets.

Kihon Happo Technique Six

As before, when your opponent grabs you, move back at a 45-degree angle. This time, however, as you rock back in, snake your arm under his arm and lock the elbow, as shown. Rotate to your left and lower your weight to toss your partner to the ground.

When attempting this technique, you may have a tendency to lose your natural body position or wrestle with the arm of your opponent. Stay naturally erect and lock that elbow in the crook of your arm. Keep your arm bent at almost a 90-degree angle with your hand up. Turn smoothly and lower your weight to take the opponent to the ground.

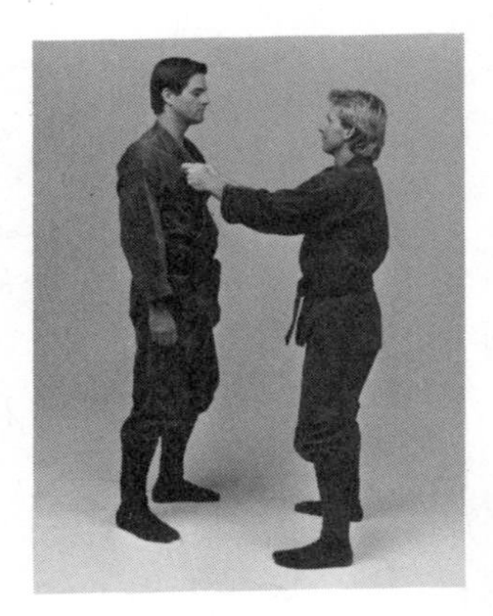

Kihon Happo Technique Seven

Your partner grabs you with both hands. Punch into your partner's right wrist as you move back to your left at a 45-degree angle. Next, place your right arm behind your partner's elbow, as shown. "Walk" forward until your hands meet, rotate to your right, and drop your weight. This action will take him to the ground where he can be immobilized.

The secret to this technique is simple. "Walk" your hands together until you can clasp them, don't use arm strength to pull them together. Keep your body naturally erect and use your weight to bring your partner down.

Kihon Happo Technique Eight

Your partner grabs your lapel with his right hand and swings to punch you in the face with his left. Simultaneously, you strike the incoming wrist, cover his hand with your right hand, as shown, and step back and to the right at a 45-degree angle. This should take your partner slightly off balance. Push his hand off your lapel with both hands and step back to the left, again, at a 45-degree angle. Rotate to your right and put your left knee on the ground to force your opponent down.

This last is one of the most basic of all of the kihon. I have complicated it slightly by adding the punch. The secrets of this technique are: distancing and angling, defeating the balance of your opponent with taijutsu, and staying out of the range of incoming strikes. This technique should require no power.

There are an infinite number of other basic techniques and variations. The above represent common ones that are taught in the Bujinkan dojo. Remember, however, that these are training exercises. In a real fight you will probably never use these exactly as shown. These are notes and scales, of which you may use bits and pieces, as you move spontaneously in a self defense situation. With practice, you may gain a degree of spontaneity and creativity as you train. Unfortunately, you cannot rehearse for a real fight. They never occur in the way that you have practiced for them. But the training *method* can prepare you for a real fight—once you recognize the difference between a fundamental exercise and a spontaneous flow based on experience gained from practicing those exercises over and over. The closest analogy remains the difference between individual notes or predetermined scales, and the spontaneous, creative flow of notes that occurs in a saxophone jazz solo. Think about it.

In the meantime, an interesting exercise will be for you to go back to each of the exercises above and try them each with an earth, then water, then fire, then wind attitude. For your void training, mix and match the techniques and feelings, or run two or more together in a series. You will have material for many months of training.

A FINAL WORD OF ENCOURAGEMENT

As I close the final major section of this book, I remind you to seek the guidance of those more experienced than you in perfecting your taijutsu. Particularly in the beginning, it is hard to recognize unnaturalness in yourself. A teacher can help you. It is difficult to find good teachers, but they are out there. I have found them, you can too.

Keep a good perspective on life as you train. Fighting is such a small part of life. A rational man wishes that it was nonexistent. Unfortunately, it is not, so the warrior personality trains to perfect his fighting arts. For your own health and sanity, I suggest that if you share my absorption in the warrior arts, that you have other interests of a more genteel nature to balance out that part of your personality.

For example, Dr. Hatsumi wishes that he was more famous as a painter than a martial artist. If you have ever seen his artwork or calligraphy, you may believe that he still might become so. I, myself, am a fanatic, yet undiscovered musician. I still practice everyday and work toward my goal of having my music heard by others.

I do not teach martial arts for a living. I am afraid that one of two things would happen: Either I would spend so much time concentrating on martial arts that I would become unbalanced, or, my love of the martial arts would become diluted.

When you think about fighting all of the time, you become sick. The only way to save yourself is by taking the teeth out of your warrior art and allowing it to become something less—such as an exercise system or a philosophical hobby without realistic use.

Remember most of all, that the other two aspects of your self: mind and heart, both depend on the body as their temple. Take care of your body, keep it strong, keep it ready for action. And use it for just and moral purposes.

This brings us to the final topic of this book: Where does one go next? To know where to go next, you need to decide what you want to know next. This is a question only you can answer. Perpetually questing spirit that I am, I want to know everything! Obviously that is not possible. I have to decide what is important for me to know next. If a person comes to me and asks what they should be working on next, I really can't tell them. They think that I am just being distant or mysterious, but I really don't know. Out of all the things that can be known in the universe, how should I know what another person needs to know next? Who is to say that I am even qualified to teach him? That is his decision. On the other hand, if a person comes to me and says: "I can't get this particular wrist twist to work, can you help me?" Then I will be glad to try.

Ultimately, the responsibility for your training is your own. I wish you the luck that I have had in finding persons of intelligence, sensitivity, and action to look to for advice, inspiration, and example. They are hard to find, but they are out there. And maybe it isn't just luck that allows me to share their knowledge.

AFTERWORD

The confluence of mind, heart, and body creates a potential for great happiness. If someone were to ask me, "Jack, what is the meaning of life?" I would answer: "Happiness." I am not perfectly happy, few people are. But my intermittent unhappiness seems to usually occur when I fail to focus myself on accepting things that I know to be true. I, too, occasionally deny what is real.

A man without a heart is a brute. A man without facility with his body cannot control his interaction with this world. A man without a brain is dead. I am not qualified to present myself as an embodiment of the perfect, balanced, heroic human being. I am here to say that *that* is what we must strive toward. The process is a happy one.

I do feel qualified to help people gain an appreciation of the value of ninpo taijutsu movement. Taijutsu is a way of flowing that cannot adequately be illustrated in a static photo sequence. Readers interested in learning ninpo taijutsu from a qualified instructor should contact:

The Warrior Information Network
3067 East Waterloo Road
Stockton, California 95205